The Japanese
and Western Science

The Japanese and Western Science

Masao Watanabe

Translated by Otto Theodor Benfey
With a Foreword by Edwin O. Reischauer
and a new Epilogue by the author

UNIVERSITY OF PENNSYLVANIA PRESS   Philadelphia

Library of Congress Cataloging-in-Publication Data

Watanabe, Masao, 1920–
    [Nihonjin to kindai kagaku. English]
    The Japanese and Western science / Masao Watanabe ; translated by
Otto Theodor Benfey ; with a foreword by Edwin O. Reischauer and a
new epilogue by the author.
        p.   cm.
    Translation of: Nihonjin to kindai kagaku.
    Includes bibliographical references and index.
    ISBN 0-8122-8252-3
    1. Science—Japan—History.   2. Science—Philosophy—History.
3. Evolution—History.   I. Title.
Q127.J3W3713   1990
509′.52—dc20   90-13039
                                                                      CIP

Contents

Foreword

Edwin O. Reischauer

Dr. Masao Watanabe's brief book entitled *The Japanese and Western Science* is a fascinating contribution to our knowledge of modern Japan. This is because it is written by a Japanese scholar for Japanese readers and deals not with the mastery of technology, at which the Japanese have proved so adept, but with the acceptance of the concepts of pure science, which borders on philosophy.

A Westerner presented with the subject "The Japanese and Western Science" would probably discuss original Japanese weaknesses in science, possible propensities as compared with other non-Western nations, earlier glimmerings of Western science through the so-called "Dutch learning," the gradual introduction of scientific concepts and achievements, skillful Japanese imitations of Western science despite the scarcity of original contributions, and the current blossoming of Japanese inventiveness as well as superb technological skills. There is very little of all this in Watanabe's book, though there is a certain amount of overlap in detail concerning a few of the Western scientists who figured in the introduction of science to Japan. Instead, Watanabe deals largely with the intellectual and philosophical problems the Japanese met with in adopting Western science and the difficulties they still encounter. For example, their relationship with nature, which the Western scholar might miss entirely, is given very full and revealing treatment.

Much of the book is devoted to Darwinism, which illustrated the differences in the way in which science was accepted in the West and in Japan. In the West, Darwinism proved a difficult hurdle to the acceptance of modern scientific ideas, but in Japan it produced no such problem. Because of Shinto ideas, there were no clear lines between natural objects, such as rocks, trees, waterfalls, and mountains, and living creatures of all sorts, vegetable or animal, and humans, or be-tween humans and gods. Buddhism had also brought the idea that the quality of one's present life might bring rebirth as a superior being or as an inferior one, like a bug or a worm. No one recoiled at the idea that humans could be descended from less advanced animal forms. In fact, Darwinism proved a support to the acceptance of Western science.

The Japanese took readily to the idea of social Darwinism. It lent support to their already ingrained belief that the superior would tri-umph over the inferior, and they embraced the concept of the survival of the fittest, taking for granted that, despite the temporary physical superiority of the West in military and economic power, Japan's superi-ority in spirit would make it eventually supreme.

Watanabe does not discuss these topics as the reverse of the picture I present here. He comes to the question in his own way, which is uniquely Japanese. To him, the Japanese adoption of science is basi-cally an exercise in philosophy. As he puts it, "when modern science is looked at as the product of a Western worldview, and particularly when its formative phase is examined, its true nature as an intellectual and spiritual human activity becomes evident."

This is a most revealing book, worthy of study by anyone who wishes to know more about modern Japan, its tremendous success in modern science, and the problems it still faces in the scientific field. It must be admitted that the Japanese are changing radically, and younger Japanese who, unlike Watanabe, received their education entirely after World War II might well display very different attitudes toward science. But this is a significant book, which will greatly reward the reader.

Translator's Preface

Masao Watanabe's *The Japanese and Western Science* was brought to my attention in Tokyo by a German visiting professor at International Christian University. He had bought *Die Japaner und die Moderne Wissenschaft*, the German translation of Watanabe's *Nihonjin to Kindai Kagaku*, and lent it to me. I was fascinated. One of my aims at ICU had been to obtain more information on how Japan had acquired Western science and how it had adjusted to it. It was hard to find any coherent information in German or English.

I searched for an English translation of Watanabe's book, but learned there was none. On the other hand, the author was nearby, living five minutes from the ICU campus, and was occasionally teaching history of science there.

Masao Watanabe was one of the creators of the Graduate Course of History and Philosophy of Science at the University of Tokyo, where he taught until reaching the university's mandatory retirement age. He later taught at Niigata University and Tokyo Denki University, and in 1988 created the History of Science Program at ICU. Over the years, he had become one of the scholars most knowledgeable about the *yatoi*, the foreign teachers brought to Japan to initiate that country's transformation into a modern industrial giant. In search of biographical and background information, he was for a time a visiting scholar at the Harvard-Yenching Institute in Cambridge, Massachusetts, where he became acquainted with Edwin Reischauer, later to become the United States ambassador to Japan.

The author and I met in 1974 at the 14th International Congress of the History of Science held in Tokyo and Kyoto. There he showed a film on the Japanese mirror described in Chapter 2 and took some of the participants to visit Yamamoto Kōryū, the last surviving practitioner of the magic-mirror craft. In 1986 Watanabe showed me his own mirror and its remarkable powers.

This book is in the form of case histories written by a Japanese for Japanese, revealing the atmosphere and trends in the sciences during the Meiji period through particular individuals and themes.

The translation process was somewhat unusual and deserves a comment. In view of my very limited mastery of the Japanese language, I made the translation from the German version and transmitted completed chapters to Masao Watanabe for checking against the original. A number of questions were clarified during his later visit to Greensboro, North Carolina, where I was teaching at Guilford College.

The Japanese way of dating historical events by giving the year of a given reign period has been converted in the translation to the more conventional system. The reign periods, occasionally referred to in the book, correspond to the following dates:

Heian   ninth to twelfth centuries
Edo   1600–1868
Meiji   1868–1912
Taishō   1912–1926
Shōwa   1926–1989

Thus 1985 was the sixtieth year of Shōwa. Japanese personal names, on the other hand, have been left in the Japanese manner, family names preceding given names, except for the names of the author, which appears in its English form everywhere except in the Bibliography.

Theodor Benfey

Acknowledgments

My translation of this book follows a long history of involvement with the Orient. While I was teaching history of science at Earlham College, Indiana, Jackson Bailey, head of the East Asian Studies program, suggested I spend a year in Japan to acquaint myself with the Chinese and Japanese contributions to my field. Nathan Sivin, now of the University of Pennsylvania's Department of History and Sociology of Science, guided my initial summer's investigation of the Oriental sciences. With a Fulbright/Hays research and study fellowship, my wife, Rachel Benfey, and I together with two of our sons, Stephen and Christopher, spent 1970–71 at Kansei Gakuin University in Nishinomiya. Tokunosuke Watanabe, Dean of the Science faculty, was my host, while Shigeru Oae of Osaka, later head of the chemistry department of Tsukuba University, helped much in our orientation, as did Eikoh Shimao now of Doshisha University. Joseph Needham's *Science and Civilization in China* became my indispensable published guide. I later had the great pleasure of visiting him and his collaborator Lu Gwei-Djen at their research center in Cambridge, England. Our second son Philip independently spent an extended period in the Orient and contributed insights for my work.

For our second full year in Japan, 1985–86, at International Christian University, my host was Koa Tasaka. To take advantage of the opportunity, William Rogers, President, and Samuel Schuman, Vice President for Academic Affairs, of Guilford College, North Carolina, allowed me to take my sabbatical a year earlier than rightfully my due. David McInnes, my department head, showed much forbearance towards my ever greater involvement in Oriental studies. Many others have been generous with their expertise—the librarians at ICU, Earlham and Guilford, scholars in Oriental science in Japan and here, and finally Arnold Thackray and his colleagues at my new location at the Beckman Center for the History of Chemistry.

In the production of the book, Patricia Smith, Rita Colanzi, Kathleen Moore, and Alison Anderson of the University of Pennsylvania Press have helped make the preparation of the manuscript a learning experience and a joy.

My thanks to all these friends and helpers.

My part in this book I dedicate to the Tokyo section of my family, Stephen Benfey, Kikue Kotani, Maiko, and Alisa.

Theodor Benfey

Introduction: Japan's Modern Century

Over a hundred years have passed since Japan began to assimilate modern science and technology from the countries of the West. The research and educational institutions created for that purpose—the University of Tokyo, founded in 1877 and Sapporo Agricultural Institute, founded in 1876, which became Hokkaido University—have just celebrated their centennial. Scientific societies such as the Mathematical Society, the Physical Society, and the Chemical Society of Japan, all founded very early, have just become one hundred years old.

In these hundred years, the Japanese turned their greatest energies to adopting an alien culture and civilization in place of their own. For a while, concentration was totally on the West; at other times attempts were made to preserve part of the national tradition under the banner "Japanese ethics, European science." From time to time the pendulum swung further in one of the two directions; in many respects a superficial mixing of the two cultures took place that is easily discernible in Japanese daily life, politics, business, industry, education, and learning. The two elements exist side by side. Often the new element is not rooted deeply enough to make much contribution. In other places, the genuine traditional values are in process of being forgotten, and at times perceptive foreigners are needed to bring their significance back to our consciousness.

This situation needs to be remedied. The next hundred years must be used to explore new possibilities, but for that purpose we need first of all to look back and survey the development of modern science in Japan from its origin to our time. This is even more relevant since all over the world people have begun to worry that modern science and technology are facing a turning point in their development.

The Power of Modern Civilization

As everyone knows, modern science and technology claim universal validity. Yet modern science began only in the sixteenth and seventeenth centuries and the technology based on it reached its height of effectiveness only in the nineteenth century. Both at first took place within the culture of the West.

When Columbus ventured to cross the Atlantic at the end of the fifteenth century we are told that the ships were out of sight of land for so long that his crew implored him to turn back for fear they would reach the "end of the world." Beyond the familiar world there lay, it was believed, a realm, the "end of the world," where accumulated knowledge and experience were not applicable.

Then in the second half of the seventeenth century Newton appeared. Through him one learned that the same laws of gravity hold on earth as in the heavens. The comets, feared until then as evil omens, came to be seen as nothing abnormal, and their paths became accurately predictable. In the sciences, at least in the realm of mechanics, knowledge has advanced so explosively that the "end of the world" has become very remote.

The methods of modern science in the meantime have also been successful in areas outside mechanics such as optics, materials science, biology, heat, motive power, electricity, and magnetism. Where the new knowledge in these areas connected with modern technology, its development accelerated so that it can be said with justice that "knowledge is power."

In Search of Knowledge

The power of modern science and technology was extraordinarily great. Not only did it expand our view and deepen our intellectual understanding of the world, but its practical application contributed to the proliferation of industrial production, to the development of transportation and communication on land and sea, and to the strengthening of national economic and military power.

Toward the middle of the nineteenth century, faced with the might of Western countries possessing science and technology, Japan was forced to open its borders. It would have been impossible for Japan to assert itself in the new international community without opening itself to the outside and actively adopting modern science and technology. The European power represented by the "black ships" of the American commodore Matthew Perry that forced Japan to open its borders, the power of Western civilization that Fukuzawa Yukichi epitomized as "steam and electricity," this remarkable power that was able to establish itself throughout the world also had to absorb that country in the distant Orient, Japan. The Japanese slogan was "Seek Western Knowledge and Strengthen the Imperial Domain." The best and fastest method to achieve this end was to send able young men to study overseas and to bring American and European teachers into Japan. Students were sent overseas first by the provincial daimyos and then by the central government. Many foreign teachers were hired; indeed, their pay sometimes corresponded to that of heads of Japanese ministries.

Thus began the introduction of modern science and, at the same time, the modernization of Japan. All this occurred at a tempo that surprised even Western countries. Over a very short period Japan underwent a tremendous transformation. Yet this is also the reason why many problems remain unsolved. The present book takes up and analyzes a number of episodes from this historical period, during which modern science was being introduced and accepted, in order to clarify and uncover the attitude of the Japanese toward the West, and thereby to discern where the major problems are located.

Some Examples

The book begins with the case of Yamagawa Kenjirō, who was sent to study in America at the beginning of the Meiji era. Yamagawa came from the Aizu province and was a very young member of the Byakkotai.* On completing his foreign studies he became the first professor of physics at the University of Tokyo, and later its president. By looking at Yamagawa's life and work we can not only obtain an insight into the general situation at the time but enter into the feelings of a Japanese, overseas for the first time, who came into contact with the ideas and culture of the West; we can see how he reacted to them and what he accepted and did not accept.

At the same time that many Japanese were sent abroad, many foreign teachers came into the country at the invitation of the Japanese government. Their activities led to various scientific explorations of specifically Japanese phenomena and objects, and they also initiated a cultural exchange between Japan and the West. A concrete example— the studies of the *makyō*, the "magic mirror"—is described in Chapter 2. Fascinated by this mysterious phenomenon, clearly a product of great craftsmanship and skill, foreign scientists proceeded to examine it. The results became known in the West and led to reactions of various kinds. Young Japanese scientists were stimulated by these researches and also began to study the *makyō*.

---

*The Byakkotai, the White Tiger Corps, were a group of young samurai from Aizu who fought against Satsuma and Chōshū and when defeated committed harakiri, ritual suicide. Being too young, Yamagawa was not permitted to join this battle.

Chapter 3 introduces E. S. Morse as representative of foreign teachers in Japan. Morse discovered the prehistoric sites of Omori shell mounds, introduced into Japan the theory of evolution, and was the first zoologist at the University of Tokyo. This case study reveals the differences in scientific level between Japan and the West in that period, as well as ways in which modern science was introduced into Japan and the problems associated with its introduction. Even more strongly than other foreign teachers, Morse was attracted to things Japanese. He studied them with great enthusiasm and introduced them to the West.

Among scientific theories brought to Japan, the theory of evolution had the most enduring effect. How was the Darwinian doctrine, which at the time not only was the most important scientific theory in the West but also shook the Western worldview, received in the intellectual world of Japan and what discussions did it set in motion? Clarifying this question helps us understand the particular form and manner in which modern science found an entry into Japan and at the same time illuminates in an important way the intellectual history of the Meiji era. Chapter 4 focuses on this theme.

In general the introduction and further development of modern science and the formulation of new theories proceeded independently of traditional Japanese culture and its spiritual background. As a particular exception, the case of Oka Asajirō considered in Chapter 5 deserves our attention. During his studies in Germany Oka acquainted himself with biology and the theory of evolution. He digested this knowledge and reflected upon it according to the framework of Japanese Buddhist views of the transience of life, arriving at original and characteristic conclusions. Oka developed a number of insights some of which even today have lost none of their cogency.

A fundamental difference is evident between the traditional Japanese conception of nature, in which nature is viewed through the eyes of a poet and where the aim is to become one with nature, and the Western conception, in which nature is a creation subordinate to the human level, and is treated as an object of scientific research and an instrument to be used. Comparing and contrasting these two conceptions make clear why modern science developed in Europe and why the West today is looking with great interest at traditional Japanese culture. At the same time we can begin to understand why modern science could not have originated in Japan and what underlies the problems that arose in connection with its introduction. Finally, these observations allow us to reconsider from the viewpoint of a Japanese such contemporary problems as conservation and environmental pollution. In Chapter 6, I share my personal insights regarding these problems in the light of my experiences, observations, and reflections.

In considering the introduction of modern science and the acceptance of Western culture in Japan, three fundamental problems can be identified:

(1) The simple technical introduction, initiation, and utilization of Western achievements, a process that ignored their philosophical and cultural background.

(2) The transfer of each of the specialized areas of knowledge separately, which took no account of their intimate interconnection with various realms of Western civilization.

(3) The acceptance of Western culture and knowledge, which paid no regard to Japanese traditions, thus making the two strands run side by side without making contact.

Correcting these problems is our cultural task today and must become a focus of a broadly and deeply based education. It is a necessity for the Japanese who, as Japanese, want to assert themselves as persons with Western knowledge in the international community. My personal reflections and perspectives regarding these questions form Chapter 7 of this book.

Chapter 1
From Samurai to Scientist: Yamagawa Kenjirō

Modern science originated in the West in the seventeenth century, but even there it did not become a professional field until the second half of the nineteenth century. Until then the only intellectual professions demanding a set of prescribed qualifications were theology, medicine, and law.

After Japan opened its doors in the second half of the nineteenth century, in order to catch up with the West it adopted Western knowledge as well as institutions in one fell swoop. Modern educational institutions were created that began to train their own research and teaching staffs. Many of their students came from samurai families, for in this area new career opportunities became possible for them after the daimyo system was abolished. One of these was Yamagawa Kenjirō of the Aizu clan, who was to become the first professor of physics at the University of Tokyo. Yamagawa contributed much to the introduction of physics into Japan and in addition made major contributions to his country as an educational statesman.

Turning Toward Western Science

Yamagawa Kenjirō was born on July 17, 1864 in Aizu-Wakamatsu, located in what is now the Fukushima prefecture. His father died young, and he was brought up by his mother and his grandfather, who was chief administrator of the Matsudaira clan. In his memoirs Yamagawa describes his grandfather as a very progressive person who early saw to it that all the members of his family were inoculated against smallpox, introduced Western rifles, and was the first in the Aizu-Shijimi clan to culture shellfish. Yamagawa's older brother Hiroshi later became a major-general, president of the Higher Normal School in Tokyo, and a member of the Upper House. He was also made a baron.

His older sister became a lady of the court, and his younger sister, later the wife of Duke Ōyama Iwao, was along with Tsuda Umeko one of the first women to study abroad. All this shows how progressive the Yamagawa family was and how conscious it was of the importance of education.

In 1862, when he was eight years old, Kenjirō entered the Nisshinkan school in Aizu-Wakamatsu. This school, which he attended for six years, had been established by the Aizu clan. Its curriculum concentrated on Confucian and samurai virtues. Neither Japanese history, geography, foreign languages, nor science was taught. Abacus (arithmetic) was available, but since it was considered to be a subject fit only for merchants and not for samurai Yamagawa did not take advantage of it.

In 1868, the first year of the Meiji era, civil war erupted between the Tokugawa shogunate and the emperor, in which the Aizu clan became enmeshed. The Nisshinkan school was closed and its fourteen- to sixteen-year-olds were organized into the now famous White Tiger Corps. Of course, Yamagawa Kenjirō, then fourteen, joined also. When the clan decided that fourteen-year-olds were not able to handle the heavy rifles of that time, he was released and ordered to learn French instead. The procedure was to copy French texts with a brush onto Japanese paper.

After the capitulation of Wakamatsu castle, Yamagawa fled to Echigo in what is now Niigata prefecture, and for a while studied Japanese and Chinese texts in the city of Niigata. In 1869 he moved to Tokyo, and the following year studied English in the Aizu clan's private school in the Zōjōji temple. This school, however, closed soon afterwards; so he switched to the Numama-juku, where he not only learned English but taught it. It was a hard life. Here he also had mathematics lessons for the first time. The man who later became professor of physics at the University of Tokyo thus learned multiplication only at sixteen. At that time he had difficulties dividing two-digit numbers, and not even in his dreams would he ever have thought of studying physics. Slowly he familiarized himself with geometry and the basic concepts of algebra.

In those days there were few schools in Tokyo where foreign texts could be studied. These were the Daigaku-Nankō [the precursor of the University of Tokyo], the Keio Gijuku of Fukuzawa Yukichi, the Dōjinsha of Nakamura Masanao in Koishikawa, as well as the Kyōkan-Gijuku, which Fukuchi Gen'ichiro had opened in Shitaya. I wanted to attend either the Daigaku-Nankō or the Keio-Gijuku, but I was poor and so was my clan and therefore my studies could not be financed.

Thus Yamagawa describes this period (posthumous manuscripts of Baron Yamagawa).

Japan's Future and Physics

The Kaitakushi (the department in the new Meiji government in charge of the development of Hokkaido) decided to send its future officials overseas for training. It was believed that the people most suited for cold Hokkaido would be those from northern provinces, so in addition to young men from the more southern Satsuma and Chō-shū clans, a few were chosen from the Aizu and Shōnai farther north. One of these was Yamagawa Kenjirō. Thus it came about that although he came from the Aizu clan sometimes despised as "foes of the emperor," and had been released from the White Tiger Corps only because of his age, and although his studies in Tokyo had been fraught with difficulties, Yamagawa received orders from the emperor's government to study overseas.

At first the government assumed that the best place for training people for opening up the northern territories would be Russia, and that the young men should be sent there. The students, however, showed little eagerness for the idea, and those who had already been overseas thought that studies in Russia would not be productive. In the end the group was sent to America. On January 1, 1871, in Yokohama, they boarded the *Japan,* the most modern paddle steamer of the time, and reached San Francisco on January 23. Yamagawa is reported to have worn strange Euro-Japanese clothing and old white shoes that were far too large.

In America he completed a year of basic training in a secondary school and then sought admission to the Sheffield Scientific School of Yale University in New Haven, Connecticut. He was accepted on condition that prior to entry he would master trigonometry. The Sheffield Scientific School was the precursor of the Science Division of Yale University. Founded in 1847 as the Yale Scientific School, it had been renamed in 1861 in honor of the industrialist Joseph Earl Sheffield, who had contributed buildings, furnishings, and finances.

The subjects taught there at the time were chemistry, civil engineering, agriculture, natural history, medicine, mining, and metallurgy. Physics as a subject was not yet offered; so Yamagawa decided to study civil engineering "in order to gain some knowledge of physics." Later he also took advanced mathematics courses. The Yale University catalog of that time lists not only the courses but also the names of teachers and students, among them the name, nationality, and New Haven address of Yamagawa Kenjirō.

Figure 1 The Sheffield Scientific School.

Figure 2 Yamagawa Kenjirō during his student years in America (presumably in 1875 at the time he was completing his studies).

Why was Yamagawa so interested in studying physics? If we select from the available documents those parts that could have a connection with his choice, one experience stands out from the voyage to America that impressed him. He described it as follows:

In those days I had not yet overcome my ingrained contempt for foreigners and I was not ready to respect them. But then something happened that impressed me greatly and led me to realize that I needed to learn from them. In the middle of the Pacific we were informed that that evening or the next morning we would meet the boat of the Pacific Mail Company and those who wanted to send mail to Japan should have it ready. I seriously doubted whether on this huge body of water two ships could possibly meet as planned. At three or four in the morning we in fact met up with the other ship and stopped about 200 meters from it. A boat was let down, our letters were taken across and others received. Watching this, I was deeply impressed by the superiority of Western knowledge and decided that in the face of such massive knowledge Japan would be impotent. (*Writings of Baron Yamagawa*, pp. 48–49)

Another clue comes from his decision at the beginning of his studies at the Sheffield Scientific School. He wrote:

Never before had I thought so intensively about my future. Just at that time Herbert Spencer published a new philosophical work that deeply impressed the younger generation. A Dr. Youmans of New York publicized Spencer's views in the magazine *Popular Scientific Monthly*. I thought about his teaching and came to the conclusion that Japan had to be brought to a position of prosperity and power, but to attain that would require an improvement in its politics. To improve its politics, society had to be improved, which requires knowledge of sociology. Sociology in turn presupposes the study of biology and the other sciences. To make real the slogan "attain prosperity and strengthen the military," physics and chemistry are required. I therefore determined to study physics. (p. 53)

The magazine *Popular Scientific Monthly* was founded by Edward Livingston Youmans (1821–1887) in 1872, the same year that Yamagawa began his studies at the Sheffield Scientific School. The purpose of the magazine was to convey awareness of the ever-increasing body of scientific knowledge to a broad public.

Youmans adopted Herbert Spencer's (1820–1903) systematic attempt to interpret all phenomena ranging from the inorganic world to human society on the basis of evolution theory. It was under the influence of these ideas that Yamagawa regarded "society" as linked with "biology and the other sciences."

Stimulated by direct contact with Western civilization, Yamagawa thought anew about the future of Japan in the world and decided that the study of physics was extraordinarily important. Dr. Nakamura Seiji and Dr. Tamamushi Bunichi, both of whom knew Yamagawa personally, report that he originally had intended to serve his country as a politician. Since, however, his origin in the Aizu clan blocked the possibility of a political career, he decided to devote himself as a scientist to research and teaching. He wanted very much to show by means of his scientific activity that the Aizu clan was no longer "an enemy of the emperor." These were the motives underlying his decision, which in turn determined his life.

After studying for a year and a half at the Sheffield Scientific School, Yamagawa suddenly was ordered to return to Japan. He refused, however. Fortunately Lucy Baldwin, the wealthy aunt of his friend Robert Morris, offered him financial assistance so that his studies could continue. Mrs. Baldwin made her assistance dependent on his signing a commitment. Yamagawa had decided to turn down the offer if it required him to accept Christianity, but he was to promise only that after completing his studies and returning to Japan he would devote all his energies to the well-being of his country. In this way the future president of the Imperial University of Tokyo, through the assistance of a generous American woman, was able to complete a three-year program of studies at the Sheffield Scientific School, obtain a bachelor's degree, and return home as a twenty-one-year-old in May 1875.

His rejection of Christianity mellowed over the years. He later wrote:

> I went overseas when I was very young. Many Japanese on the basis of traditional conceptions held that Christianity was wrong and I too was determined to hold to this, which I now think was a mistake. Thus, on principle, I could not enter any foreign church. At Yale, where I studied, everyone went each morning to the university church to pray but I did not feel free to participate. I, therefore, obtained permission from the administration, not only at Yale but throughout my stay overseas, never to set foot in a church. When I think about it now I realize I did not have to go that far, but, as I have said, at the time I considered Christianity mistaken. Not once in those five years did I enter a church. (*Writings of Baron Yamagawa*, pp. 253–54)

Yamagawa wrote this as advice for students embarking on a tour abroad, advising them as far as possible to adapt themselves to the customs of the country they were visiting.

The Beginnings of Physics

In 1876, the year after his return from America, Yamagawa was appointed assistant professor in the Tokyo Kaisei School. He taught there together with Peter Vrooman Veeder (1825–1896), an American physics teacher, with whom he cooperated in giving lectures and carrying out experiments. The following April the Tokyo Kaisei School and the Medical School of Tokyo were combined to form the University of Tokyo. Yamagawa became assistant professor in its College of Natural Science. At the same time he continued to assist in Veeder's classes.

When the College of Natural Science was first established in the University of Tokyo, the scientific lectures were mainly given by foreign teachers, but soon Japanese teachers took their places. Yamagawa was appointed a full professor in July 1879, becoming the first Japanese professor of physics in Japan. He taught together with the American Thomas Corwin Mendenhall (1841–1924), who had come to Japan the previous year as Veeder's successor. After Mendenhall left in 1881, the subject was taught only by Yamagawa and the Englishman James Alfred Ewing (1855–1935), who stayed until 1883. In 1881 he married the second daughter of Niwa Arata of the Karatsu clan. In April 1888 he was named a doctor of science, becoming one of the first twenty-five Japanese to receive the doctorate.

Yamagawa's teaching ranged over the whole field of physics. After he was joined by Tanakadate Aikitsu, who had completed his physics studies at the University of Tokyo in 1882, and later by others, Yamagawa was able to divide the course work so that he could concentrate mainly on theoretical physics. In later years he worked in the fields of optics, thermodynamics, acoustics, and capillary phenomena. During this period he was the first in Japan to light an arc lamp by means of Bunsen batteries. The event took place in a public demonstration at a graduation ceremony at the university in December 1877. At a similar ceremony in July 1882 he succeeded in lighting an Edison incandescent lamp by means of a dynamo.

Yamagawa's scientific studies are mostly limited to the period 1875–1890; thus with few exceptions they all occurred before his thirty-seventh year. In later years he concentrated on educational activities. The most important experiments, publications, reports, translations, and books published by him are described here.

(1) Researches on the *makyō* in 1880 (see Chapter 2), though he published nothing on this subject.

(2) Japanese translation of Mendenhall's "Report on the Meteorology of Tokyo for the Year 1879," published in the journal *Rikakai Sui* in 1880 and 1881. Mendenhall at the time was also head of the University of Tokyo's meteorological station.

(3) "Fires in Tokyo," published in *Rikakai Sui* in 1881 and simultaneously in English in the journal *Memoirs of the Science Department,*

University of Tokyo. On the basis of descriptions of major fires in the Edo period, he attempted to find a connection between the number of fires and wind velocity, as well as the number of typhoons and connections between the directions of fires, winds, and typhoons.

(4) "A Brief Introduction to Spectroscopic Methods," published in 1885 in the journal *Proceedings of the Physico-Mathematical Society of Tokyo.* This paper deals with the description and establishment of the following phenomenon: When the solar spectrum is observed with a spectroscope, both ends of the spectrum become diffuse because of the dispersion of light by the prism. This can be avoided by the use of suitable colored glass.

(5) "A New Method for Measuring Capillary Constants" describes a method using a spring balance for determining the capillary constant. The complete text was written in the classical Japanese style but with Romanized Japanese and was published in 1886 in the *Proceedings of the Physico-Mathematical Society of Tokyo.*

(6) "Determination of the Coefficient of Thermal Conductivity of Marble" describes a procedure for measuring the coefficient of thermal conductivity of a marble sphere and the results of such measurements. The article was published in 1889 in English in the *Journal of the College of Science, University of Tokyo.* A summary, also in English, appeared in 1888 in the *Proceedings of the Physico-Mathematical Society of Tokyo.* A summary in Japanese appeared in *Journal of the College of Science, University of Tokyo* in 1893. Yamagawa's results were included in the German *Handbuch der Physik* in 1896.

(7) *Dictionary of Physics Terminology, Japanese, English, French, and German,* published in 1888 in collaboration with Muraoka Han'ichi. At Yamagawa's request, more than thirty scientists worked several years on this eighty-eight-page physics dictionary. It is in four parts, each in alphabetical order: Japanese-English-French-German, English-Japanese-French-German, French-Japanese-English-German, and German-Japanese-English-French.

(8) X-ray studies (1896), carried out within a few months of Röntgen's discovery of X-rays. Kuwaki Ayao writes in his *Notes on the History of Science (Kagakushi Kō)* that Yamagawa was the first Japanese to experiment with X-rays. The experiments were reported in articles in the magazines *Tōyō Gakugei Zasshi* and *Tokyo Butsuri-Gakkō Zasshi.* Although Yamagawa lectured about them, he seems not to have published anything about them himself.

(9) "Experiments Regarding the Clairvoyance of Mrs. Nagao" (1911), attempts to study extrasensory perception experimentally. At that time it not only attracted general attention but even interested

Figure 3 Dictionary of Physics Terminology, Japanese, English, French and German.

scientists who wanted to investigate the claims by scientifically controlled experiments. This is the only example of scientific research by Yamagawa in his later years.

(10) Other work included a report of the proceedings of an international conference on electrical units (1885), a review of a foreign book on electricity (1886), lectures on electromagnetism (1888), and others.

None of these researches indicates significant originality. Yet they are the kinds of projects that had to be carried out by a physicist in a country like Japan needing to overcome in the shortest possible time a major lead by the West. The compilation of the *Dictionary of Physics* was in this respect a particularly important task. The study of Tokyo fires supplied a basis for city planning and fire prevention. Yamagawa's data on the coefficient of thermal conductivity of marble were internationally recognized. However, the most conspicuous of Yamagawa's studies were those on X-rays and extrasensory perception.

Experiments with X-rays and on Extrasensory Perception

Almost as soon as Röntgen's discovery of X-rays was reported, Yamagawa, with Assistant Professor Tsuruta Kenji, began his experiments. By studying the differences between the effects of X-rays and cathode rays on magnetism and on photographic plates, they confirmed Röntgen's views. Yamagawa carried out public demonstrations and lectured on the subject. Kuwaki Ayao describes Yamagawa's experiments as follows:

The creation of the department of physics of the University of Tokyo in 1877 occurred after a period when major discoveries were being made in Europe. Cathode rays, discovered by Crookes, were examined. Röntgen's discovery occurred at the end of 1895 and was immediately picked up in Japan. Dr. Yamagawa was the first Japanese to carry out such experiments. Since no X-ray tubes were available in Japan at the time (they were introduced in the latter part of 1896) he made them himself according to Sprengel's directions. Marconi's discovery occurred the following year. (Kuwaki Ayao, *Notes on the History of Science*, pp. 191ff)

In 1895, immediately after Röntgen's discovery, Professor Yamagawa and Assistant Professor Tsuruta Kenji attempted to repeat the experiments with inadequate instruments. With much effort they were at last successful. A medical journal reported that he demonstrated them to the Medical Society. No details were given but according to his assistant at the time, Mizuki Tomojirō, Yamagawa on this occasion, among the experiments he conducted, tried also the transmission of X-rays through crystals. Had conditions been more favorable, Yamagawa might have anticipated Max von Laue." (pp. 508ff)

In the journal *Tōyō Gakugei Zasshi* there is a report at that time by Nagaoka Hantaro, who was studying in Berlin, on "The X-ray Studies of Mr. Röntgen." In addition to the description of Yamagawa's experiments there is also an article on the X-ray studies of Mizuno Bin'nojō, a professor at Daiichi Kōtō Gakkō, a high school in Tokyo. The explosive development of physics in Europe thus was quickly transmitted to Japan.

In his later years Yamagawa devoted himself mainly to educational administration and seldom took part in research. Only once did he undertake a piece of research, the so-called "clairvoyance" experiments.

In April 1910 there appeared in Kumamoto a woman, Mifune Chizuko, with reportedly clairvoyant powers, who attracted much public attention and stimulated the interest of scientists. On September 14 of the same year, experiments with her were carried out in Tokyo in the presence of the scientists Yamagawa Kenjirō, Oka Asajirō, Tanakadate

Aikitsu, medical doctors Kure Shūzō, Ōsawa Kenji, Katayama Kuni-
yoshi, Irisawa Tatsukichi, and Miyake Hiidzu, and doctor of philosophy Inoue Tetsujirō. In November of that year in Marugame a woman,
Nagao Ikuko, claimed not only to be clairvoyant but to be able to
discern the taste of hidden foods and to develop any letter or character
assigned to her on photographic plates by her mental power of sensitiz-
ing the plates. Miura Kōsuke, a psychology student at Kyoto University, traveled to Marugame to investigate. He concluded that a special
light effect was involved that he termed *Kyōdai kōsen* (Kyoto University
rays). He also published these conclusions. Dr. Fukurai Yūkichi, psy-
chologist at Tokyo Imperial University, termed the effect *nensha* (mental imaging). Because of Fukurai's intense interest in clairvoyance he
had already studied Mifune Chizuko, and he now took up experiments
with Nagao Ikuko of Marugame. In those days, cases of clairvoyance
were reported from many places, relating to shipwrecks, the search for
drowning victims, or the solving of crimes.

Yamagawa, who was then president of the Meiji Semmon Gakkō
(Meiji Professional School) in Tobata on the island of Kyushu, asked
Tanakadate in October to prepare materials for an experiment and
sent them to Kumamoto and Marugame, but both women failed to
demonstrate their power. However, since by December the wave of
clairvoyance had spread throughout Japan, Yamagawa had the head of
the Girls' High School in Marugame announce a direct experiment. In
January 1911, he stopped in Marugame on a trip to Kyushu to examine the case himself. Instructor Fuji Noriatsu and doctoral candidate
Fujiwara Sakuhei (both physicists) participated. Materials for the investigation were carefully prepared. But through a series of mishaps, the
film material was lost and, although it was sought for even by the
police, it was never located. Consequently, no definite conclusions
could be reached. It is clear that repeated suspicious occurrences connected with these experiments increased the doubts of scientists regarding clairvoyance claims.

From the fall of 1910 to the following summer, scholarly journals
such as *Tōyō Gakugei Zasshi* and *Rigakukai* were full of articles on clairvoyance. The great interest of scientists, especially of physicists, was
not altogether without reason. In the first place there were the epoch-
making discoveries in physics in the second half of the nineteenth century, especially the discovery of electromagnetic waves (Hertz, 1888),
wireless telegraphy (Marconi, 1895), X-rays (Röntgen, 1895), radioactivity (Becquerel, 1896), and radium and polonium (Marie and Pierre
Curie, 1898). The existence of a number of invisible effects and rays
had thus been detected. For that reason the interest of physicists in
unseen phenomena and in claims of clairvoyance was intense. In fact,

they used the same procedures for studying clairvoyance that they had used for X-rays and radioactivity. They presented the objects to be "viewed" not wrapped in paper or in an ordinary container but enclosed in lead so that X-rays and radioactivity could not penetrate it. Yamagawa was, as was mentioned, the first Japanese to study X-rays. Ono Suminosuke (instructor at the Daigo Kōtō Gakkō, the fifth high school of Kumamoto, and eventually a professor at the Tokyo Institute of Science), who later joined in the experiments, had some experience with radioactivity.

The second reason has to do with the social significance of this phenomenon. Fukurai Yūkichi, who was interested in it from the beginning, and Miura Kōsuke, who later became interested, were both psychologists. They respected clairvoyance as such and judged it positively, and thus the public at large was prone to believe in it. Such a phenomenon, however, needed cautious tests for support, and thus it had become necessary to test its authenticity by strictly scientific means. Yamagawa succinctly describes the situation at the time:

The general public talks simply of "clairvoyance studies" but at first it was necessary to find out whether this mysterious phenomenon, as generally assumed, actually existed. If it turned out that it did, its properties had to be discovered.

To answer the first part of the question—whether there is such a thing as clairvoyance—is in my view a task for physicists, while to answer the second part, on the other hand, as to what its characteristics are, physiologists, psychologists, philosophers, etc. may be more appropriate. As is well known, I used to be a physicist and it could be that my suggestion that the question should first be examined from the point of view of physics might therefore cause some amusement. So be it. ("Experiments Regarding the Clairvoyance of Mrs. Nagao")

This is part of the introductory section of "Clairvoyance Experiments" (February 1911) edited by Fuji Noriatsu and Fujiwara Sakuhei.

A third reason might be mentioned. Yamagawa's interest in clairvoyance dated back to his stay in America, where, as he mentioned immediately after his return home, he had witnessed clairvoyance experiments at the Sheffield Scientific School with a student who had become known for his extraordinary perceptive abilities. This left a lasting impression on Yamagawa.

The Promotion of Science and the Development of Education

The center of gravity of Yamagawa's activities from the end of the 1880s slowly shifted to educational administration. In 1893 he became dean of the College of Science (corresponding today to the dean of the Natural Science Division), and in 1901, at age forty-seven, he became president of Tokyo Imperial University. From 1886 on he had also been a member of the national Education Council and became its chairman in 1900. Starting in 1897 he was, in addition, a member of the Higher Education Conference and the commission to determine the curriculum of the middle school. He became vice-president of the conference in 1906. In 1905 he resigned his post as president of Tokyo Imperial University and withdrew from public office for a few years. During this time he wrote numerous articles on education, science, and politics for the Tokyo *Nichinichi Shimbun*.

In March 1903, Yamagawa drafted a statement explaining his resignation from the National Academy and handed it with a brief explanation to the Academy's president. In the statement he justified his resignation by the fact that since becoming a university president he had had no time for the pursuit of science. To stay on as a member would not be in line with the Academy's aims, but would only block entry for younger scientists. This offer to resign from the Academy witnesses not only to Yamagawa's integrity but also to his clear intention to transfer his activities to educational administration.

In 1907 Meiji Semmon Gakkō (Meiji Professional School) was founded in Tobata in northern Kyushu with the financial help of the industrialist Yasukawa Keiichirō. It offered four-year curricula in three subjects—mining, metallurgy, and mechanical engineering. Yamagawa was called to be its president. Always eager to expand education and particularly facilities for scientific training, Yamagawa was so pleased with the aims of this industrialist that he spontaneously accepted. This school, not only through its curriculum but also in its financing by a perceptive industrialist, showed striking parallels with his American alma mater, the Sheffield Scientific School. The similarity may have been the determining factor in his acceptance. Following the common practice overseas, Yamagawa wanted to name the school after its benefactor, but Yasukawa absolutely refused and the school was therefore named Meiji Semmon Gakkō. In an almost autobiographical letter of June 1910 to the Alumni Club of the Sheffield Scientific School, Yamagawa expressed his high regard for Yasukawa for making an extraordinary contribution to his country by founding this school.

Yamagawa, who persistently advocated education in the sciences, was worried about the (to his mind) too strongly intellectual component in the new educational guidelines. He saw his task at the newly founded Meiji Semmon Gakkō as not only in the area of technical scientific and engineering training of students but also in their moral education. This was Yamagawa's central concern and it was also the wish of Yasukawa. Yamagawa's intentions are also clearly revealed in a speech he gave at the provisional opening celebration on April 1, 1909. In her drive to catch up with the West, he claimed, Japan had attracted mainly people with specialized one-sided abilities and had neglected moral education. Recently the negative consequences of this tendency had come to light. The Meiji Semmon Gakkō was therefore to be an institution for the training not only of specialists but of "men of honor who have technical knowledge."

From the time Yamagawa had chosen to go into the sciences as a young man, science and technology had been inextricably intermeshed in his mind with the problem of the state. Having only too clearly experienced at first hand the superiority of Western civilization, he was convinced that Japan had to excel in science to become rich and strong. The following is taken from a popular scientific lecture he gave to the Physico-Mathematical Society of Tokyo in 1907:

Our national debt currently amounts to 2 billion yen. To repay it we must develop agriculture and industry. Yet both agriculture and industry are based on science. Our ancestors have already said that scholars concern themselves with key questions. They show us the way. To attempt to develop industry and technology without the furtherance of science would be like trying to catch fish from trees. ("Our National Debt," *Gakujutsu Tsūzoku Kōenshū [Popular Scientific Lectures]* (1907): 1–16, p. 3)

It was in these terms that Yamagawa saw the significance of science for the needs of his country. Behind his work for school education and general public education, as well as underlying his stress on military training, lay one and the same motive. On the one hand, he proclaimed the great significance of science and advocated more schools and re- search centers, the improvement of the teaching of English, abolition of the teaching of Chinese classics, the simplification of the Japanese writing system taught in school, the introduction of Romanized Japa- nese, and the adoption of the metric system. At the same time he stressed moral education and took part in the movement for general public education, for which purpose he undertook a lecture tour throughout Japan. All these projects were motivated by the one wish, to achieve the prosperity of his country. His guiding moral principle was the patriotism of the samurai.

In this connection we should not overlook the fact that Yamagawa was not an anti-foreign militarist, but called for international justice, understanding, and freedom. In particular he was convinced that un- derstanding and cultural exchange between Japan and America was important for world peace. He himself had been able to continue his studies through the help of an American woman, and in a speech to the employees' club of the Meiji Semmon Gakkō on October 6, 1926, he mentioned that a photograph of this woman hung in his reception room. By 1880 he had founded the Aizu Scholarship Society for the education of young people, and this too is almost certainly traceable to his experience with Lucy Baldwin.

For Yamagawa, for whom the state was the most important concern, science and education had to serve the state also. For that reason he "did not find it necessary to educate both sexes equally." "For women, high school education is enough." The task of women in his view was marriage and the education of many capable and industrious children: this was the duty of women to the state.

In April 1911, Yamagawa was named the first president of Kyushu Imperial University. In 1913 he returned to his post as president of Tokyo Imperial University, a position that he held until 1920, except for a brief period from 1914 to 1915 as President of Kyoto Imperial University. In 1927 he became consultant to the newly founded Institute of Physical and Chemical Research. He himself had advocated the creation of this institute and was a major force in its founding. Several times he was urged to accept its leadership, but each time he declined. In addition, from 1919 until his retirement in 1920 as president of the University of Tokyo, he was the acting president of the university's Aeronautical Research Institute. He had for long been interested in airplanes.

In 1926 he became principal of the Musashi High School. The conditions that he demanded from the school's benefactor, Nezu Kaichirō, included a certain freedom in the commitment of finances, the introduction of a pension system, and support for overseas study leaves by the faculty. Health reasons led to his retirement in March 1931, and on June 26 his active life ended at the age of seventy-seven.

Japanese Traditions and Western Skill

Yamagawa was a scientist, educator, and educational statesman with the typical Meiji-era ideology, a combination of Japanese tradition and Western skill. His character can be seen in his phrase "There are many Japanese who have died out of loyalty, but I know of no one who has died for the truth." In later years his habit was to drink a limited amount of Japanese sake for reasons of health. In order to save rice (Japan's main agricultural product in those days), he used sake synthetically produced at the Institute of Physical and Chemical Research, measuring it out exactly with a metric graduate. It is significant also that in his address at an occasion to celebrate his seventy-seventh birthday (counted Japanese style, by which at birth you are one year old; the Japanese consider the seventy-seventh birthday of special significance), he cited statistics about old people from a demographic study, and that on another occasion he explained to his students in detail the construction and capabilities of a zeppelin airship, adding a comment about its possible military uses. His contemporaries called him "General Nogi in civilian clothes." An American teacher at the University of Tokyo called him "true and brave."

Reviewing Yamagawa's life, we see that since he was of the Aizu clan he could not take up a political career. In contrast to the leading politicians, who tended to dislike foreign influences, the Aizu clan always was in favor of contact with the outside world and the opening of Japan toward it. Yamagawa adopted this attitude. Thus, he went overseas, became a physicist, and made every effort as professor of physics and as an educational statesman with a global perspective to guide his country in the proper direction for development. In this way he succeeded in fulfilling his original wish and actualizing fully the promise that as a student he had given his American benefactress.

Yamagawa's physical researches do not seem particularly original, but this was not to be expected. His particular contribution lay in the fact that, as a physicist in the first phase of that era, he vigorously absorbed Western science in the shortest possible time, built up the requisite educational and research system, rapidly ended Japan's dependence on foreign researchers and books, and guided Japanese science to independence. This achievement deserves our fullest respect.

Yet for Yamagawa science was synonymous with science for the state. His conception of progress in science had the same basis as, for instance, his call to introduce military training into the curriculum. His nationalism was not directed against foreign countries nor was he using it for personal ends, but it cannot be denied that the views he followed regarding science and education, formulated as they were during the Meiji era, had certain limitations. They were effective in rapidly establishing and expanding a system of scientific training and research. On the other hand, they contributed nothing to the education of women. It is here that we must look for the limitations of Yamagawa's introduction of modern science and Western ideas.

The problem has to do with the fact that Yamagawa only concerned himself intensively with one of the two central elements of Western civilization, namely modern science, while he ignored the other element, Christianity. This was not only true for Yamagawa but was also more or less the position of the leading men of that time, and it helped determine the manner in which the adoption of Western culture and the modernization of the country was carried out. Today we can see that this limited outlook prevented a comprehensive understanding of Western civilization and therefore made impossible a fundamental understanding of modern science as the outcome of human intellectual and spiritual activity. Yet the historical situation in Japan at that time allowed no other development. The theory of evolution, to be discussed later, which was becoming important in the West at the time, was also responsible for this outcome. In this connection, the American teacher E. S. Morse, the subject of Chapter 3, played a role.

Against this background, another kind of development at that time in Japan gains in significance. Christian higher education (including that of women) was begun at various places while the Sapporo Agricultural Institute, a national collegiate institution in the northern island of Hokkaido, with W. S. Clark as its first president, promoted both Christian and scientific education. Here is not the place to discuss this further, yet the roles such institutions played and had to play deserve special investigation and recognition.

Chapter 2
Japan Studies of Foreign Teachers in Japan: Investigations of the Magic Mirror

Having just opened its doors, Japan was, for the West, an unknown world. Foreign teachers came there full of curiosity and with high expectations. Frequently what they saw excited their interest. Many of the scientists discovered objects and phenomena characteristic of this part of the world and proceeded to study them. There were physical phenomena such as climate, the force of gravity, and earthquakes; other areas to investigate such as Japanese fauna, flora, natural history, prehistoric shell mounds, fossils, and aspects of ethnology. The foreign scientists also became interested in Japanese intellectual history and culture and expanded their research interests to these fields. Thus they not only transmitted modern science to the Japanese but also played a significant role in international cultural exchange by introducing Japanese culture to the world at large. The *makyō* studies are one part, though only a small one, in this undertaking.

The *makyō*, or "magic mirror," is a special kind of bronze mirror that had been known in Japan for a long time. Its front is amalgamated, and even detailed examination fails to distinguish it from an ordinary mirror. If, however, the mirror is used to reflect sunlight onto a white surface, a Buddha figure or a sacred Buddhist text appears, corresponding to a design cast on the back of the mirror.

In China and Japan since ancient times mirrors have been revered as sacred objects. This particularly mysterious object was held in special awe. The magic mirror first appears in Japan in the Edo period, and the Japanese skill in crafts saw to it that at the beginning of the Meiji era numerous examples existed. They attracted the attention of foreign teachers, and thus this mysterious phenomenon was for the first time scientifically studied. It was the foreigners who called the *makyō* the "magic mirror."

Among mirrors of this kind there were some with a double layer on the back. Over the first layer with the design poured in during casting, an additional metallic layer with a different figure was applied. The result for the viewer was an extraordinary phenomenon in which the image projected by the mirror was entirely different from the one visible on the back. Among these mirrors a few were said to have projected the image of a crucifix or of the Virgin Mary and were intended for use by Japanese Christians, in hiding because their religion was banned.

In China the *makyō* had always been known as the "light-transmitting mirror" because it seemed as if the light went through the mirror to portray the figure on the back. That is a clever designation but it cannot be quite right because the brighter parts of the image correspond to the thicker parts of the mirror while the darker parts correspond to the thinner.

Robert William Atkinson: The Theory of Varying Coefficients of Reflection

Robert William Atkinson (1850–1929) was the first to study the *makyō* scientifically. Atkinson was an English chemist who had come to the Tokyo Kaisei Gakkō, later the University of Tokyo, as a chemistry teacher in 1874. He carried out a number of researches in Japan, among them a study of the fermentation of rice wine. The English scientific journal *Nature* on May 24, 1877, published a brief report by him about the *makyō*.

In this report Atkinson expressed the opinion that the origin of the phenomenon is to be sought in the variable pressures occurring during the filing of the mirrors. He found through his own experiments that a scratch made on the back causes the corresponding location on the front of the mirror to produce a brighter reflection. Because of the picture cast on the back, the mirror is of variable thickness, and hence variable pressures are exerted during filing. Atkinson concluded from this that there are corresponding variations in the coefficients of reflection so that the thicker parts of the mirror reflect more light (or, differently stated, absorb less light) than the thinner ones. In this way the pattern on the back is reproduced on the projection surface as light and dark variations.

In the second half of his article, Atkinson cites a part of a report given by Anton Johannes Geerts (1843–1883) to the Asiatic Society of Japan (published in *Transactions of the Asiatic Society of Japan*, vol. 4) on Japanese metals and metallurgy. The quoted section deals with Japanese bronze mirrors. In it Geerts describes the composition, the

Figure 4 A *makyō* demonstration.

Figure 5 Robert William Atkinson.

method of casting and filing, the mercury amalgamation processes on the mirror surface, and its polishing, but not the *makyō* phenomenon.

William Edward Ayrton and John Perry: The Theory of Varying Curvature Due to the *Megebō*

Atkinson's report contained only simple experiments and suggestions. The researches of William Edward Ayrton and John Perry, on the other hand, were of a more basic character. Ayrton (1857–1908) taught telegraphy and natural science from 1873 to 1878 at the Technical College, which was later merged into the Imperial University of Tokyo. Perry (1850–1920) taught physics and mathematics at the same college. Both were British. In 1878 they published a joint article entitled "The Magic Mirror of Japan" in the *Proceedings of the Royal Society* (December 12, 1878, pp. 127–48).

The article's introduction cites three reasons for the interest of foreigners in these mirrors: (1) they are of special significance in Shinto shrines; (2) Japanese families consider them as objects of considerable value; and (3) they have the unusual property of reflecting the raised design on the mirror's back. It was the third reason that interested Ayrton and Perry.

In their article they mention that one of them (probably Ayrton) had learned about the Japanese *makyō* from Charles Wheatstone before leaving England in 1873. Both Wheatstone's explanation and one by David Brewster in the *Philosophical Magazine* of 1832 spoke of simply scratching the design on the surface of the mirror and then polishing it, so interest in it was limited. When he came to Japan, however, he was amazed to discover that even the mirror dealers had no explanation for their amazing properties. Ayrton and Perry say in their report:

> We have since learnt, however, by diligent inquiry, that as is the case with many things appertaining to Japan, so with the magic mirror, the people who know least about the subject are the Japanese themselves, and we think this only furnishes another proof that teachers to instruct the Japanese about Japan itself are the greatest *desideratum*. ("The Magic Mirror of Japan," p. 129)

They then discovered Atkinson's article already referred to..

Following this introduction, Ayrton and Perry's article gives a brief overview of what past research had discovered about the *makyō*, noting that Chinese and Japanese *makyō* were already known overseas. Usually, they remarked, the phenomenon was explained as due to varying pressure conditions during the formation of the mirrors and consequent variations in their metallic properties, an explanation Perry and Ayrton knew to be wrong since the Japanese mirrors were not pressed

Figure 6 William Edward Ayrton.   *Figure 7* John Perry.

in a mold but poured. In 1866 an explanation appeared in the journal *The Reader* by a person named Parnell, who argued that the phenomenon arose from the fact that, during cooling after pouring the radius of curvature of the thinner parts of the *makyō* is greater than that at the thicker ones. This too Ayrton and Perry doubted on the basis of their own results.

Ayrton and Perry also found that except for a single document, the *Shinhen Kamakura-shi*, there was practically no Japanese literature on the *makyō*. That document described a valuable old mirror in the Kenchōji temple in Kamakura that shows on its back an ocean scene, a plum tree, and the shadow of a sickle moon; when the mirror is viewed at an angle, a Buddha figure appears. The mirror for that reason is an object of particular veneration. Ayrton and Perry report that another document, the *Kokon Itō*, describes a method for applying a picture on the front surface of a mirror by means of a special glue that is unaffected by the polishing process. Ayrton and Perry hence suspected that the Kamakura mirror was treated in this way. In their article they then proceed to report experiments for producing a *makyō* in line with the method of a Tokyo mirror maker, after which they consider possible explanations for the magic mirror phenomenon, listing them in the following classification:

1. The pattern might be scratched on the face of the mirror and hidden by subsequent polishing.

2. The portion of the face corresponding with the pattern might have a different molecular constitution from the metal forming the remainder of the mirror.

This difference in molecular constitution might produce the results:-

    a. By causing the portion of the face corresponding with the pattern at the back to attract more mercury, and so to become capable of being polished more easily; or

    b. By causing it to be harder, and so to acquire a better polish; or

    c. By causing it to polarize light.

This difference in molecular constitution might be produced:--

    a. By the inlaying of another metal; or

    b. By portions of the surface being acted on chemically; or

    c. By unequal density produced by inequality in the rate of cooling. . . .

3. The phenomenon might arise from the face of the mirror having intentional or accidental inequalities on its surface, in consequence of which, the part corresponding with the pattern on the back might be relatively concave, and so concentrate the light, or, at any rate, might disperse it less than the remainder of the slightly convex mirror. (p. 136)

The author's next step is to test these points experimentally. On the basis of experiments with a polarizer they discard the possibility of light polarization. From optical experiments regarding the curvature of the mirror surface, they conclude that the third hypothesis, the variation of curvature, is the correct one. The description of these experiments is very detailed. Parallel and converging light rays were reflected from each part of the mirror, the reflections were analyzed, and hence the varying curvature of the surface deduced (see Figure 8).

How did the varying degrees of curvature arise? Ayrton and Perry investigate further. By studying a large number of bronze mirrors they find that the design on the back of thick mirrors fails to be projected and that the surface curvature is slight. Then the authors examine the materials used for making the mirrors and the method of production, especially the casting. They make the remarkable discovery that the inside of the casting mold is almost flat while the surface of the mirror normally is curved. From this they conclude that the curvature does not arise from the casting but from a later process, the use of the *megebō*, as follows. After the mirror is cast, the surface of the mirror is roughly filed. It is then placed on a wooden board with the mirror surface up, and the mirror maker files the mirror in one direction using an iron rod rounded at one end and about half an inch in diameter and one foot long. This is the *megebō*, or "distorting rod." This filing motion makes the surface of the mirror curve cylindrically, the axis of the cylinder being parallel to the filing scratches. Next the *megebō* is used at right angles to the original direction, and then in directions in between (some mirror makers apply the *megebō* spirally) until the whole surface arches uniformly upwards. The back of the mirror is therefore curved concavely in the process. (This, however, is only possible with thin mirrors.) The mirror maker then uses a hand-scraping tool to polish away the scratches produced in the filing.

The use of the *megebō* causes the upper surface of the mirror to become generally convex, yet those parts that correspond to the figure on the back and that are therefore thicker show a lower curvature; that is, their radius of curvature is greater. Hence arise the varying degrees of curvature on the mirror surface corresponding to the design on the back and causing variations in light reflection. This is Ayrton and Perry's explanation of the *makyō* phenomenon. They hold it as possible that part of the curving of the mirror surface might also occur during the last phase of mirror production. They therefore end their account with a description of the final polishing and the coating of the surface with mercury amalgam.

Ayrton left Japan in June 1878. On January 24, 1879, he gave a lecture at the Royal Institution in London about the Japanese *makyō*, which was published in *Nature* the same year. In the lecture, Ayrton

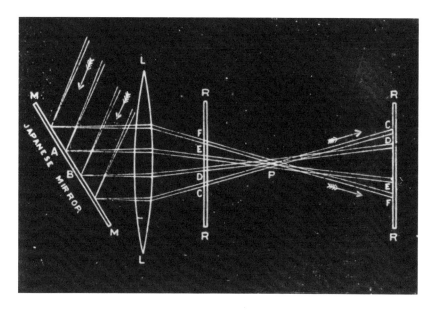

Figure 8 Optical experiment for the determination of the curvature of a mirror surface.

explains the reason for the veneration of mirrors in Japan in connection with the "three imperial insignia" (mirror, sword, and jewel) and other myths. He then describes his and Perry's experiments. The lecture's contents are essentially identical with the description given here.

Gustave E. Berson: The Theory of Variation in Curvature Due to Thermal Stress

Gustave E. Berson was probably the first to publish a scientific study of the *makyō* in Japan. Berson was French, had lived in Japan since 1876, and taught physics at the University of Tokyo. In 1880 he published in the journal *Gakugei Shirin* an article in Japanese, "The Strange Properties of Japanese Bronze Mirrors." Ono Kiyoteru made the Japanese translation, and Samejima Susumu revised it. Berson introduced his report with an overview of the *makyō* literature in the West and the Orient, then presented Ayrton and Perry's research results, and followed with his own experiments aimed at the quantitative testing of their conclusions.

In order to determine quantitatively the radius of curvature of the various parts of the mirror surface Berson used the method of Newton's rings. He was able to confirm that the surface of the mirror is normally convex and that the radius of curvature at thick parts of the mirror is greater than at thinner parts. Thus, for instance, it was 5.1 m at a thick place while at a thinner one it was 1.7 m. Using Newton's rings, he also found unexpectedly that when he made a scratch in the back of a mirror with a nail a concave line was produced on the front. He also measured the change in curvature of the surface produced by pressure on the back.

In his article, Berson proposed a hypothesis for the change in curvature as a cause of the *makyō* phenomenon. His argument, which probably sounds much more complicated in the Japanese translation than in the original, suggested that, as the bronze solidified after pouring, an internal stress was created in the metal. This stress caused differing radii of curvature on the surface while the mirror was being polished, depending on the thickness of the material.

Magic-Mirror Studies by Japanese Scientists

Stimulated by the studies of foreign scientists, Japanese physicists such as Gotō Makita and Muraoka Han'ichi also began to publish research on the *makyō*. Yamagawa Kenjirō had already (in 1880) carried out some *makyō* experiments, but he never published his results.

Gotō Makita was a student of Fukuzawa Yukichi. He studied at the

Keio Gijuku University and later taught at the Normal School in To-kyo. Stimulated by Berson's research, he studied the *makyō* together with Mihashi Tokuzō and Yamaji Ichiyū and published a two-part article in 1883 in the journal *Tōyō Gakugei Zasshi.*

This article, entitled "Investigations of Japanese Bronze Mirrors," is also written in not-easily-understood Japanese. Its results can be summarized as follows:

(1) Mirrors and all kinds of coins reflect what is on their back if they are polished thin enough.

(2) Unevennesses on the mirror surface of a thin bronze mirror occur on polishing because of the unevennesses on the mir-ror's back and corresponding to the thickness.

(3) Additional observations.

These experiments were not very systematic and had the disad-vantage that no quantitative measurements were carried out regarding salient questions. Although the Japanese scientists refer to Berson's article in *Gakugei Shirin,* they mention neither Atkinson's report nor Ayrton and Perry's studies in foreign journals. Probably they had not read them. Later, in 1887, Gotō Makita spent some time studying in England.

Muraoka Han'ichi graduated from the Daigaku Nankō, later the University of Tokyo. There he learned German, and after doing educa-tional work in the Ministry of Education and the Normal School for Women began his studies in 1881 at the German University of Strass-burg. He studied with Kundt and Röntgen and returned to Japan with a diploma. In 1882 he was appointed professor of physics in the medical school of the University of Tokyo and began intensive re-search, including his *makyō* studies.

The first articles Muraoka published on this topic appeared in 1882 and 1883 in the journal *Tōyō Gakugei Zasshi* and were entitled "The Reflected Images of Vibrating Mirrors and the Sand Figures on Their Surface" and "Comments on the Images Reflected from Vibrat-ing Mirrors." These, however, did not go to the heart of the *makyō* phenomenon but rather had a preliminary character. Muraoka studied the surface of a mirror by using a violin bow on its edge to bring it into vibration and observing the reflected sunlight thrown on a white wall. He compared this method for observing surface vibrations with the Chladni sand pattern method.*

---

*Ernst Florenz Friedrich Chladni (1756–1827) devoted much of his life to the study of acoustics and vibrations. The technique named for him involved sprinkling sand evenly on the surface, and studying the pattern produced by running a violin bow over the surface edge.

Figure 9 Muraoka Han'ichi.

His 1883/1884 article in *Tōyō Gakugei Zasshi* entitled "Explanation of the Makyō" is a summary of his researches on the cause of the *makyō* phenomenon. He observed the production of the *makyō* and found experimentally that the surface of bronze plates arches up when scratched or polished, particularly in the case of thin plates. This leads to the conclusion that the place on the surface of a mirror directly above a protruding location on the back must be concave. Hence it must reflect light more strongly than other parts. According to Muraoka, this is the cause of the *makyō* phenomenon. He suggests that the arching of the surface from scratches or polishing is due to molecular action, but he undertook no studies to confirm this hypothesis. Nowhere in "Explanation of the *Makyō*" is there a description of apparatus for measuring the unevennesses in the metal's surface, and Muraoka used no procedures for determining their radii of curvature. To a large extent, his explanations are the same as those of Ayrton and Perry, though lacking the comprehensiveness and precision of the latter.

In 1884, a German translation of Muraoka's article appeared in the *Annalen der Physik und Chemie*. Later, in 1886, he published a paper in German in the *Journal of the College of Science, University of Tokyo*, as well as in the *Annalen*, on the mathematical analysis of the elastic deformation of metal plates on grinding.

The *makyō* studies described here are a clear case study of the state of science in its beginning stages in Japan. Foreign teachers were the first to interest themselves in the *makyō* and to attempt a scientific explanation. Stimulated by them, the first generation of Japanese scientists also took up this topic. In comparison with the accounts of the foreigners, a certain immaturity, seen particularly in the absence of quantitative and concrete data, non sequiturs in the arguments, and logical looseness, cannot be overlooked. Often their experiments were not systematically designed, and the experimental conditions were often not sufficiently defined or quantitatively established. In many cases, the kind of experiment under discussion is unclear, the process of experimentation cannot be deduced from the accounts, or data on the measuring procedures used are lacking. Any stated aim of elucidating the phenomenon quantitatively is absent also. In addition there was the difficulty of having to formulate a scientific argument or an experimental description in the written Japanese in use at the time. All this witnesses to a situation in which the Japanese were not yet familiar with scientific patterns of thought or research methods.

By learning from the foreign teachers, however, they slowly became familiar with scientific procedures and experimental methods. Thus, in the *makyō* studies they moved on to investigations of plates of other metals. The discovery that a mirror of any metal exhibits the *makyō* phenomenon when ground thin enough can be listed as their independent achievement.

The clarification of the *makyō* phenomenon was by no means completed. But the topic was not a central topic for science, it was not dealt with further, and scientific interest in it slowly became dormant.

The Echo in America: The Student Annual *The Makio*

In addition to those already mentioned one other foreign teacher studied the *makyō*—namely, Thomas Corwin Mendenhall, the American physics teacher at the University of Tokyo, who had come to Japan in 1878. In the documents of the time there are reports about him, but since he himself did not publish his studies, we know nothing about them. He was recommended to the University by Edward Sylvester Morse, the American zoologist, who was already teaching there.

Morse had returned to America in 1879 and lived in Salem, near Boston. In the spring of 1880 he was invited by Ohio State University to present a lecture. Mendenhall taught there both before and after his Japan stay, but in 1880 he was still in Japan. Morse's lecture was called "Things Japanese." At the University of Tokyo, Morse had often visited Mendenhall's laboratory. Probably he became acquainted with the *makyō* there for in his lecture he also told about "magic mirrors." Students at Ohio State were in the process of launching a student annual. The lecture gave them the idea of using the Japanese word for magic mirror as their title. They wrote to Morse, who had by that time returned home, asking him for the Japanese word and its written characters. Morse immediately located a Japanese person living in Boston, confirmed that the word "makio" was the correct one, had the characters drawn with brush and ink, and sent them to the students.

Figure 10 Cover of the first issue of the annual *The Makio*, published in 1880.

Thus it came about that this publication from its first issue in June 1880 carried the official title *The Makio*. The brush-stroke characters sent by Morse graced the cover. The salutation of this issue read: "We desire that, like the real mirror, it may reflect from the surface the image of our outer college life, and under strong sunlight on close inspection may reveal the hidden picture of inner college life." X-rays to illuminate the inside had not yet been discovered.

This annual of Ohio State University continues to be published today under the same name. The letter Morse wrote to the students at the time is preserved in the university archives. Today, however, hardly anyone knows the meaning and reason for the title *The Makio*.

The Magic Mirror in Subsequent Years

In 1908, Mori Sōnosuke, professor at the Daisan Kōtō Gakkō (Third High School) in Kyoto, published a textbook called *Experimental and Theoretical Physics: Optics*. The third chapter, entitled "The Japanese Makyō," includes a short account of Muraoka Han'ichi's studies. Muraoka, who at the time was professor in the Science and Technology faculty of the Kyoto Imperial University, had himself checked the book.

Twenty-five years later, the famous English physicist and Nobel prize winner William Henry Bragg (1862–1942), who became known for his X-ray diffraction studies, wrote three pages on the *makyō* in the last section of the introductory chapter, "The Properties of Light," in his textbook on optics, *The Universe of Light* (London, 1933). Bragg described Ayrton and Perry's results and included a picture of *makyō* experiments. He ended with the following comment:

It seems very strange that so slight a departure from the regularity of the surface should produce an effect so obvious. But perhaps we may compare what we see here with other observations that we have made. . . . Perhaps we have noticed the ripples on the ceiling when water outside the window . . . is stirred by a very slight breeze and reflects the sunshine . . . yet the ripples themselves are not at all easy to see when we look for them on the surface of the water. (W. H. Bragg, *The Universe of Light* [London, 1933], p. 37)

The *makyō* also surfaces in Japanese literature. In the first part of the 1926 short story "The Hell of Mirrors" by Edogawa Rampo we read:

Thinking about it now, he seemed to have a strange passion for objects that portray others, such as glasses, lenses, and mirrors. Proof of this is that his toys consisted almost completely of such objects as a projector, a monocular telescope, and a lens. . . .

And I remember the following from his childhood. When I once visited him in his study, there was an old wooden box on the table, and he was holding an old bronze mirror that probably was kept in the box. He was using it to throw sunlight onto a dark wall of the room.

"Isn't that interesting? Look! When I throw light on the wall with this flat mirror, a strange written character is seen there."

When I looked at the wall as asked I found to my amazement that in the circle of light, slightly deformed to be sure, the character for "luck" in bright golden white light shone back! "How strange! How is that possible?"

It seemed miraculous. On me, still a child, it had a strange, a frightening effect.

"You don't understand this. Shall I explain it to you? It's really quite simple. Look at the back of the mirror! See here in relief is the character for luck. That shines through to the surface."

As he said, there was in fact a superb relief pattern on the back of the bronze-colored mirror. But why did it shine through to the surface and throw such a picture on the wall? The mirror surface was perfectly smooth no matter from what angle one looked, and one's face in the mirror was not distorted at all. Only when the mirror was used to reflect light could one see the strange image. It seemed magical to me.

"This is not magic." He saw my unbelieving face and began to explain. "My father told me that a metal mirror, unlike a glass mirror, has to be polished occasionally because otherwise it slowly loses its shininess. This mirror has for long been in the family's possession and has often been polished. But every time it is polished, a little more metal is removed from those parts opposite the design on the back. This comes from the fact that where the metal is thicker the pressure when polishing is greater also. This minute difference, invisible to the eye, causes the design to be reflected by the mirror. Do you understand?"

Admittedly I came to understand the cause from this explanation, but the idea that a face can mirror itself clearly in this surface even though it shows unquestionable unevennesses when reflecting light, this mysterious fact was as uncanny to me as the feeling when one looks into a microscope and sees tiny little things there. I shuddered. (Edogawa Rampo, "The Hell of Mirrors," 1926)

A reader, already familiar with the *makyō* will find this passage of interest. Without prior information, however, one can hardly make much sense of it. This shows that even in reading literature one needs knowledge of science and its history. This is particularly true of Western literature.

For a case study of the history of the introduction of Western science in Japan, I devoted myself intensively to the history of Japanese *makyō* studies. From my investigations arose the idea of making a film about it. This film, *Makyo,* was produced by the Nirenoki Kōbō company in Akishima, Tokyo. Although focusing on the limited topic of the *makyō*, it gives at the same time many insights into science, art, crafts, history, sociology, intellectual history, and international cultural ex-

Figure 11 Polishing a mirror surface (from the film *Makyō*).

change. The music, in traditional Japanese style, was specially composed for the film and incorporated into it.

The Kyoto mirror maker Yamamoto Kōryū, officially designated one of Japan's "living treasures," is probably the only one in Japan today to carry on the traditional Japanese mirror-making art. The second part of the film, which shows the Yamamoto family involved in each stage of the *makyō*-making process, is particularly impressive and is a valuable historical documentary.

The film has also appeared in English. Immediately after its completion it was shown to the delegates of the 14th International Congress of the History of Science in Kyoto and Tokyo in August 1974 and was received with great enthusiasm. Some of the delegates visited Yamamoto's workshop; others wanted to obtain a *makyō* and a copy of the film. After about a hundred years, the *makyō* thus has once again aroused the interest of foreigners.

Is then the strange *makyō* phenomenon now fully understood? Not completely, I think. Even in the last twenty years Japanese and foreign scientists have examined the *makyō*, and their conclusions do not always agree. If we summarize them, we can identify three causes of the *makyō* phenomenon:

(1) The mirror surface is overall slightly convex. Depending on the thickness of the mirror the radius of curvature varies (the thicker parts are flatter, sometimes even concave). Two reasons for the origin of this unevenness can be advanced:
(a) differences in pressure during grinding; and
(b) residual thermal stresses from casting.

(2) After casting, the rate of cooling depends on the thickness of the material. Accordingly the structure of the metal becomes inhomogeneous, producing non-uniform surface reflection.

(3) If a chisel is used on the back of the mirror, the places on the front corresponding to the scratches produced become deeper with polishing.

Although I have not seen any, it is known that there are magic mirrors of type 3. The factors mentioned under point 2 may possibly be true, but it is hardly likely that such clear reflected images can be formed in this way as found with the usual *makyō*. I myself do not know whether a *makyō* corresponding to 2 ever existed. On the other hand, we can be sure that the factors given in point 1 apply to all the *makyō* described in this chapter, including the newer ones produced by Yamamoto Kōryū. The hypotheses of Ayrton and Perry, Berson, and Muraoka Han'ichi all agree in tracing the *makyō* phenomenon back to the surface form of the mirror.

By means of Figure 12, point 1 can be detailed further. When sunlight or other parallel light rays are reflected from a thin-walled place on the mirror, the rays are dispersed and throw only a weak light on the wall because the surface is convex. On the other hand, parallel rays falling on a thick-walled region are reflected almost parallel or are even bunched together (where the mirror is somewhat concave), so that the light intensity is increased. In this way a dark and light image is formed on the wall corresponding to the varying thickness of the mirror due to the relief picture on the back.

On the basis of the following simple observations, I was able to confirm the correctness of the first thesis:

- The reflected figure is slightly larger than the figure on the back of the mirror. This is to be expected if the general curvature of the mirror is slightly convex.
- To obtain a clear image, a distance from the wall must be chosen that is characteristic of the given mirror. This distance is different for each mirror, since each has a different concavity. There are, however, mirrors for which the distance on the wall

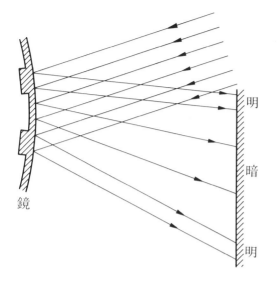

Figure 12 Deriving the *makyō* phenomenon from the nature of the mirror surface (words in figure: mirror, light, dark, light).

hardly influences the image's sharpness. In these cases the places on the mirror corresponding to the design on the back are essentially flat.

- If one looks at a straight object in the mirror (rather than projecting it), the straight lines appear wavy at those places lying opposite the design on the back. This too shows that the radius of curvature is slightly variable.
- If a glass disk, ground flat and of good quality, is placed on the mirror, Newton's rings are produced corresponding to the variation in radius of curvature.

The question of the origin of the variable curvature is probably correctly answered by explanation 1(*a*) above—namely, the variations in pressure during filing. It is quite conceivable that the second explanation, residual thermal stress from casting, plays a secondary role. The *makyō* phenomenon, however, would take place even without this factor—that is, due solely to variations in pressure while grinding.

Yet even explanation 1(*a*) does not completely solve the problem. During grinding, what kind of pressure is exerted on what part of the mirror and what effect and change in the metal takes place? How exactly do the thin and thick parts of the mirror differ? This must be clarified for each phase of the process. Although Ayrton and Perry, Gotō Makita, and Muraoka Han'ichi to a certain extent focused on these questions and formulated hypotheses, these issues have not been conclusively researched and await future clarification.

At the same time we should not forget that *makyō* of type 3 actually exist and that type 2 might exist also. Here, too, exact analyses of what goes on during their production are lacking.

Finally it should be mentioned that, as Ayrton and Perry indicated, there are mirrors of quite a different type whose surfaces are treated with chemicals or by carving or inlay. They may be called *makyō* in a broader sense.

Chapter 3
A Zoologist Fascinated by Japan: Edward Sylvester Morse

The foreigners who came to Japan in the beginning of the Meiji era came with the aim, more or less, of changing Japan. This was either imposed on them by their missionary calling or expected of them by the Japanese government, which had invited them for that purpose. Initially, however, Edward Sylvester Morse belonged to neither of these groups. Morse had come to Japan to pursue his research specialty, the brachiopods, for Japan was particularly rich in these species. Mostly by chance he was called to be professor of zoology at the University of Tokyo, and through his discovery of the shell mounds at Omori and his introduction into Japan of the theory of evolution made a name for himself in that country.

Thus it turned out that he too undertook to change Japan, but in the process he was molded by Japan to a large extent. He was so strongly fascinated by it that neither in his daily life nor in his researches could he continue without Japan. Aside from zoology, he gave much time and energy to studying Japan and spent the rest of his life as a Japanophile in America. This chapter, devoted to him, describes a foreign teacher who changed and was changed by Japan.

From Shellfish to Ceramics: A Many-Sided Life

Edward Sylvester Morse was born June 18, 1838, the son of a fur trader, in Portland, Maine, an American port city. From childhood he was always full of curiosity, and it is said that he often stole out of the classroom to wander about in the fields and on the shore. In those days, in the port cities of New England, it was the fashion to collect seashells as decorations. Sailors brought rare seashells home and scientists were busy classifying them. Young Morse too was interested in shells, which he eagerly collected and sorted. He also joined the Portland Society of

Figure 13 Edward Sylvester Morse (1863).

Natural History. Here he became acquainted with a number of natural-ists and was encouraged to do more intensive studies. His field was land shells in Maine. He published his findings and began correspondence with scholars in the field. His first article dealt with a new species of snail, and in 1857, when Morse was nineteen, it appeared in the *Proceedings of the Boston Society of Natural History.*

Morse first became technical draftsman for a railroad company, but Professor Jean Louis Rodolphe Agassiz, who appreciated his abil-ities, called him to Harvard and made him his assistant in 1859. Thus, during his assistantship, he was able for three years to study with Agassiz at the Lawrence Scientific School. There were other student assistants working under this world-famous Swiss zoologist, and all later became leading zoologists in America.

Morse earned money to cover his expenses with lectures and scientific drawings of shells for conchologists and at the same time continued his studies. He was a gifted illustrator and lecturer and was able, for instance, to use both hands simultaneously in making draw-ings on the blackboard. In America at that time, when cultural lectures were very much in vogue, his appearances were particularly popular.

In 1866, Morse and three other students of Agassiz, who at the time had some disagreements with him, moved to the Essex Research Institute in Salem. The following year they moved to the Peabody Scientific Academy (later the Peabody Museum) in Salem, founded through the generosity of George Peabody, and organized the American Society of Naturalists as well as the journal *The American Naturalist,* the first scientific journal for this research field in America.

His time in Salem—he was around thirty—was, as far as zoology was concerned, scientifically his most productive. In 1863 he married and built a house, and in 1870 his first son was born. The following year he obtained a degree from Bowdoin College and became professor of zoology there (until 1874). He also gave lectures at Maine State College and undertook popular-science lecture tours.

From the time of his first article on brachiopods in 1862 he had specialized in this field. When he gave a lecture in San Francisco in the spring of 1874, he learned that Japan was particularly rich in these species, so he decided to go there. To raise enough money he wrote a textbook, the *First Book of Zoölogy,* which was published in 1875. It contained only 190 pages, and the extensive field of zoology was presented without citing other scientists, based solely on Morse's own observations and views. The 158 illustrations were all drawn by Morse. The second edition appeared in 1877, as did a German translation (*Anfangsgründe der allgemeinen Zoologie,* Stuttgart). The second German edition, published in Berlin, came out in 1881. In Japan the book was translated by Yatabe Ryōkichi and appeared in 1888.

Morse arrived in Yokohama in 1877. Everything was totally new to him. A few days after his arrival he took what was then the only train in eastern Japan from Yokohama to Tokyo. Looking out of the train window, he discovered the prehistoric shell mounds of Omori. Later he gained permission to excavate and study them.

Morse began collecting in Enoshima, where there were particularly plentiful shell areas. A Japanese there asked him to teach zoology at the just established University of Tokyo. Morse accepted. The following two years, except for five months of preparation in America, were filled with lectures at the University of Tokyo, research travel, and public lectures on the theory of evolution.

The influence exerted by Morse on Japanese intellectuals by his presentation of the theory of evolution was enormous, but he in turn was deeply transformed by Japanese culture. As a result he even turned away from zoology from time to time devoting himself to collecting and investigating Japanese pottery and ethnographic materials as well as studying Japanese architecture. In retrospect Morse wrote in 1902:

Gradually I was drawn away from my zoölogical work, into archaeological investigations, by the alluring problem of the ethnic affinities of the Japanese race. The fascinating character of Japanese art led to a study, first of the prehistoric and early pottery of the Japanese, and then to the collection and

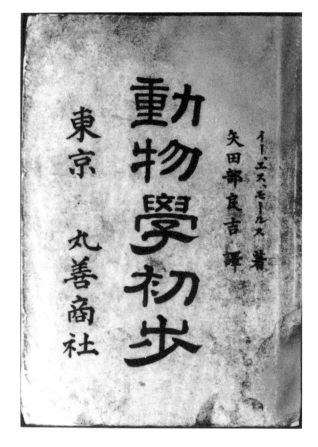

Figure 14 Cover of the Japanese translation of the *First Book of Zoology.*

study of the fictile art of Japan. Inexorable fate finally entangled me for twenty years in a minute study of Japanese pottery. (*Memoirs of the Boston Society for Natural History* 5, 8 [1902]: 313)

Later his biographer, Merill E. Champion, wrote: "Morse went to Japan to study brachiopods and returned a world authority on Japanese pottery" (*Occasional Papers on Mollusks* 1, 2 [1947]: 134). He suggested that this was a good example of what he called serendipity, the ability to make important discoveries without looking for them.

It was Morse, too, who proposed as teachers for the University of Tokyo such men as Mendenhall, Charles Otis Whitman, and Ernest

Francisco Fenollosa. After completing his contract with the University of Tokyo, he returned to America in 1879. The following year he became director of the Peabody Museum in Salem, a post he held until 1916. After retiring, he remained honorary director until his death. During this time he visited Japan once more in 1882 to collect ceramics, returning home via China and Europe. After that he refused to visit Japan again. He could not bear to see how the old Japan had changed.

Today, a bronze lantern that Morse brought with him from Japan still stands in the garden of Morse's house in Salem, where he died on December 12, 1925. His collection of Japanese ceramics is now in the Boston Museum of Fine Arts, and his folkcraft materials were donated to the Peabody Museum, where they can still be seen today. Both are first-rate collections. His publications include, in addition to works on biology and on Japan, studies on archaeology, anthropology, ethnology, architecture, ceramics and porcelain, painting, archery, music, ancient coins, astronomy, utilization of solar energy, environmental pollution (especially noise) in cities, education, museums, the history of biology, and biographies.

Figure 15 Morse's laboratory at the Enoshima coast (from *Japan Day by Day*).

Figure 16 The biology laboratory at the University of Tokyo (from *Japan Day by Day*).

Figure 17 Morse's house in Salem (author's photograph).

Morse's Contributions to Biology

Many of Morse's American colleagues pursued zoology mainly in conjunction with the theory of evolution. They investigated the various animal and fossil species of the American continent, criticizing the thesis of natural selection or confirming the theory of evolution through discovery of systematic relationships among the different species. Morse was one of them. He specialized mainly in the brachiopods, whose place in evolution he determined. In this connection he also studied the mollusks and annelids. In addition he examined the tarsi of birds and found a systematic connection between them and those of reptiles and mammals. This is Morse's most significant contribution in the field of zoology, but he also worked on protective coloring in animals and on the light of fireflies.

The jointless brachiopod species *lingula* appeared at the start of the paleozoic in the Precambrian era and reached the zenith of its development in the Silurian and Devonian. Since then it has been decimated, but it still exists, hardly having changed since that time, so that one can speak of a living fossil. Agassiz, who rejected Darwin's theory of evolution, advanced as a counterargument the surprising continuity of form of this species. Probably it was through Agassiz that Morse was led to this species, but the latter, interested as he was in the theory of evolution, could hardly have overlooked it. The classification of brachiopods was in any case a particularly difficult problem. Originally they were classified as mollusks or as a subgroup of these, as pseudo-mollusks, but the relations between brachiopods and mollusks threw doubt on that. Morse aimed with his brachiopod research to find the correct classification.

In 1862 he published his first article on the brachiopods, in which he established the fact that the two shells are not found on the sides but at the front and rear of the animal. The brachiopods thus cannot be counted among the mollusks but are more likely to be related to the polyzoas and bryozoas. Morse pursued his brachiopod studies intensively and reached conclusions confirming his hypothesis. In 1870 he published two articles, "Position of the Brachiopoda in the Animal Kingdom" and "The Brachiopoda, a Division of the Annelida," which were considered sensational among contemporary specialists. He documented that brachiopods were arthromeric spineless animals, that they were not mollusks, that they belonged to the worms close to *vermes*, and that they formed a group related to the annelida. Morse sent the article to Darwin and received the following reply:

I have just read [your essay] with the greatest interest, and you seem to me (though I am not a competent judge) to make out with remarkable clearness an

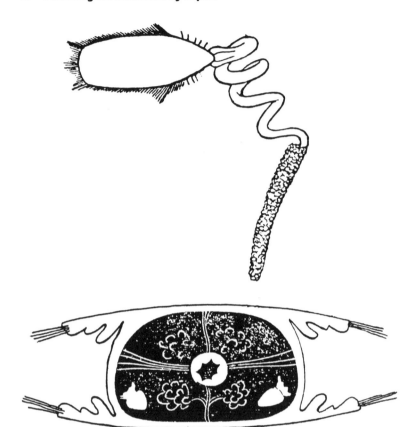

Figure 18 Brachiopods, exterior (above) and in cross-section (below) (from one of Morse's papers).

extremely strong case. What a wonderful change it is to an old naturalist to have to look at these "shells" as "worms." (Francis Darwin, ed., *More Letters of Charles Darwin* [1903], vol. 1, p. 350)

In 1873 Morse wrote one more sixty-page article on "The Systematic Position of the Brachiopoda" in which he summarized the results of his research. There he established even more clearly his thesis that the brachiopods were worms. Morse was facing increasingly vocal criticism that he relied only on the similarities in the organs of brachiopods and worms and did not consider sufficiently the differences between them. But no one is likely to disagree with the following paragraph from the review of his article in *The American Naturalist:*

Aside from the great interest of the memoir, the skilful and concise manner in which the facts,—many discovered by the author himself after the most patient study, which would in themselves commend the work to every one—are presented, we think the author has demonstrated, in the clearest manner, that the Brachiopods are worms. (*The American Naturalist* 8 [1874]: 52–53)

Morse in fact was equipped with extraordinary observational gifts and great artistic talent. These skills were of great use to him not only in his zoological studies but in others as well.

Brachiopods are not considered of much importance today, but in those days even famous zoologists took an interest in them. The significance of Morse's work lies in the classification of this species within the framework of evolution theory. And, as mentioned, it was to pursue brachiopod studies that Morse came to Japan.

Today questions are being raised about the systematics of vertebrates from the viewpoint of molecular biology. In Morse's time, comparative anatomy was the most important approach, and the attempt was made to explain the skeletal structure of the different species systematically on the basis of the theory of evolution. It had just been discovered that Mesozoic dinosaurs were related to birds, and Morse was intrigued by this thesis. If there is a connection between lizards and birds, it was strange that birds lacked (or had not been seen to have) an important group of bones that were common to lizards and mammals. This train of thought was the origin of his plan to develop a chart that, from the front and rear extremities of vertebrates that had five radiating digits at each extremity, was to explain and document their systematic relations.

For this purpose Morse took ten different bird species, among them lower forms such as the ostrich, penguin, and waterfowl, and carefully studied their front and hind limbs. He published his results in several articles between 1871 and 1880. He pointed out the difficulty in correlating the rear extremities of birds with those of other vertebrates because several of the tarsal bones were fused together. In the bird embryo, however, the metatarsal bone is still clearly recognizable. Thus he showed that the calf and shinbone were fused at the base with part of the tarsal bones. In this way, through the identification of the metatarsal bone, he succeeded in clarifying the systematic relationship between reptiles and birds. As in his studies of the brachiopods, here too his aim was above all that of systematic classification.

The Theory of Evolution

The lecture series on evolution theory presented by Morse in Japan was edited and published in 1883 by his student Ishikawa Chiyomatsu under the title *Dōbutsu Shinkaron* (*The Evolution of Animals*). In this book Morse's argumentation seems quite rough. It becomes apparent that he links causes and effects of evolution in mental leaps and applies examples from the animal and plant world indiscriminately to humans. It has generally been assumed—and until recently I shared this view— that Morse, who had studied under the declared Darwin foe Agassiz, had had no opportunity to become thoroughly familiar with the theory of evolution, or else that in Japan he could deal with the topic more superficially because opposition to the theory of evolution was hardly to be expected. Comparing his work with the state of American biology at that time in order to look into this question more carefully shows, however, that his knowledge of evolution theory was not so primitive or dilettantish as has been thought.

Morse has begun his scientific career as a student of Agassiz and his leaning toward the theory of evolution was strengthened by the group of Agassiz's students. Although Agassiz was an opponent of Darwin, his zoology prepared the ground for this viewpoint. Morse himself commented:

It has been repeatedly said, and with truth, that Agassiz's teachings paved the way for the prompt acceptance of the theory of evolution—first, because he familiarized the great public with a structural knowledge of the animal kingdom and the affinities existing between the different groups, and, second, because he demonstrated the recapitulation theory of von Baer, and added the great conception that the history of the animal kingdom from the earliest geological horizons added further proof of these principles. (*Popular Science Monthly* 71 [1907]: 546)

Darwin's *On the Origin of Species* appeared in 1859. Later, American biologists began to question the theory of natural selection. Darwin declared that among the individuals of a species random variations occurred and that through inheritance natural selection took place. However, comparative breeding experiments over several generations did not always agree with the theory. In addition, with the progress in paleontology in the second half of the nineteenth century, it became evident that in the case of fossils, a linear, directed development, that is, orthogenesis, takes place, and this too Darwin's selection theory could not explain.

Many of the American defenders of the theory of evolution were paleontologists and geologists. While Darwin emphasized natural se- lection and assumed variation, they emphasized the question of how variation occurred. They were thus critical of Darwinism and took the

Figure 19 Cover of the first edition of *Dōbutsu Shinkaron*
(*The Evolution of Animals*).

view that living organisms could also change through the influence of
the environment. This position was the so-called neo-Lamarckism.
The term itself was coined by Agassiz's student and a colleague of
Morse, Alpheus Spring Packard, Jr., and was first officially used in a
report by Alpheus Hyatt, also a colleague of Morse, in 1866, during the
period when they had both separated from Agassiz and were pursuing
their research in Salem. Yet Morse, although one of them, continued to
defend the theory of natural selection because it seemed able to explain
the many-sided adaptation of animals.

In Salem, Agassiz's students had started the journal *The American
Naturalist* in 1867. It brought out ever more articles on questions of
animal development, criticism of Darwin, and so on. Darwin referred

to these researches in *The Descent of Man,* which appeared in 1871. In America the echo generated by this book was extraordinary. It kindled the fury of theologians, who saw it as a frontal attack on the dignity of humanity. A religious person would shudder on reading Darwin's description of the ancestor of the human race as a

'hairy quadruped, furnished with a tail and pointed ears, probably arboreal in habits.' (Quoted in Richard Hofstadter, *Social Darwinism in American Thought* [1955], p. 25)

Earlier, in chapter 13 of his *Origin of Species,* Darwin had emphasized that a genuine classification had to be based on descent and, in order to be true to nature, had to be systematic. Until then, classifications had mainly noted the differences between the species. Now, however, the scientific focus was placed on the inner connections between species. Scholars concentrated on Archaeopteryx as a missing link between birds and reptiles or Amphioxus as placed between vertebrates and invertebrates, and sought to explain them in line with the system. From this grew the interest in tracing the origins of the human species.

In 1876, before the American Association for the Advancement of Science, Morse presented a long lecture with the title "What American Zoologists Have Done for Evolution." He devoted a third of his lecture to human origins. Morse at the time was vice-president of Section B of the Association and had already made a name for himself in zoology through his brachiopod research. The lecture took place a year before his voyage to Japan. It gives an insight into the status of evolution theory in America and at the same time shows Morse's attitude to the whole of zoology at the time he came to Japan.

Morse introduced his lecture by pointing to the significance that an overview of the American contribution to selection theory reveals. He divided the development of American zoology into two phases, the dividing line being Agassiz's arrival in America. The first phase was characterized by collection of material, the second by the developments that grew out of the first. He described in detail Agassiz's role in the second phase in that his earnest protest against the evolution theory led American zoologists to feel the need to search everywhere for proofs for the theory instead of simply accepting it. Needless to say, the new continent of America was a treasure trove of biological research materials, including fossils.

Further, Morse pointed out that the first clear premonition of the theory of natural selection had come from America in the first half of the nineteenth century and that Darwin had recognized this. As premises of Darwin's theory he listed the variation of species, the inheritance of particular characteristics, the survival of only a few from the large numbers of individuals, and the assumption that the natural appearance of the earth even up to our time has been constantly changing. Central to his theory was natural selection, that is, the survival of those individuals best adapted to their environment. Morse declared that the most important problem was the question whether each species as a whole has inherent in it a drive for change irrespective of its environment or whether its transformation occurs only through external stimuli. This was nothing but the question of Darwinism versus neo-Lamarckism.

In this way Morse steered the subject to the species question. The existence of species in nature is unquestioned; major support for the theory of evolution and against the theory of special creation, which claims that each species originated independently, is the fact that from species new species emerge, or that they become extinct through changes in the environment. Admittedly there were cases such as the *lingula,* a species of brachiopod, which had hardly changed from earliest times, and for Agassiz this had been a decisive counterargument against the theory of evolution. Morse, on the other hand, saw in it rather the *lingula*'s extraordinary vitality and power of survival.

Next Morse presented the researches of Packard and Asa Gray, spoke of the relationship between the geographic distribution of plants and animals and underlying geological conditions, and developed the basis for his rejection of special creation. He then described his own studies on the protective coloring of animals and demonstrated by means of examples that the behavior of animals is expressed differently depending on the environment. He thus disproved the idea that animals have implanted in them from the start a given repertoire of behavior. He referred also to insects, particularly butterflies and their seasonally determined forms, and pointed to changes in function of animals such as the disappearance of gills and the loss of sight.

Morse asked his listeners to pay particular attention to the fact that animals had been discovered that had characteristics of two or three widely separated species. Through the fossil finds of a cross between a horse and a lower form, or Archaeopteryx lying between birds and reptiles, and their systematic interpretation by Othniel Charles Marsh, Joseph Leidy, and others, the missing connecting link had been found and evolution theory secured. If then nature is uniform, is it not conceivable that the origin of the human species can be reconstructed in a similar way? In this way Morse made the transition to the problem of *homo sapiens*. He developed an explanation of the origin of human beings in line with evolution theory using research results from paleontology, anthropology, comparative anatomy, genetics, linguistics, and sociology. He cited Jeffries Wyman's studies in comparative anatomy in detail. In connection with sociology he recalled the ideas of John Fiske, a scholar of the Spencer school.

This, in rough outline, was the content of Morse's lecture. In 1887, on the occasion of relinquishing his office as president of the American Association for the Advancement of Science, he gave another lecture on the same theme. In it he presented a comprehensive overview of the further development of American zoology as expressed in journals and monographs. In contrast with his earlier lecture, he here presented the studies on heredity and clearly expressed his own views on science, religion, and society.

In his lecture, Morse also discussed Darwin's theory of pangenesis, finding support for it in a theory eleven years earlier by the American William Keith Brooks, which had been well summarized in the latter's more recently published book *The Laws of Heredity*. Morse recommended Brooks's book as the most important American contribution to Darwinism.

Regarding relations to religion, Morse commented that, although the church had been attacking Darwin's theory, educated people should feel confident that they could accept the conclusions reached by science as valid, while they should reject ecclesiastical interpretation of natural phenomena as false. Morse mentions *The Warfare of Science and Theology* by the former president of Cornell University, Andrew Dickson White, as giving many supporting examples.

Near the end of the lecture Morse discussed the responsibility of scientists for the intellectual and moral development of society. Wherever it is possible to discover by the methods of science the reasons for social grievances, it is the duty of scientists to do so. He also expressed his firm conviction that people who wanted to reform the world had no business doing so unless they knew the principles of natural selection.

Here it becomes clear in what respects Morse was concerned about society. In his later article in *The Popular Scientific Monthly* of 1892 entitled "Natural Selection and Crime," which deals with the reform of society and the principle of natural selection, he becomes even more explicit, proposing that the principle not only be used in animal breeding but be applied to humans also. Criminals and work-shy persons should be subjected to legally controlled natural selection.

These are Morse's views on evolution theory before and after his stay in Japan.

Involvement with Japan

Wherever Morse went he took a notebook with him, and everything that drew his attention he made notes on and sketched. He was a very careful observer and a skilled artist. This gift of his, which was at one time called having a "camera eye," blossomed superbly in his studies and in writings introducing aspects of Japanese culture such as ceramics, architecture, and ethnology. Thus the zoologist Morse, who initially had received no training in this field, became one of the most eminent pioneers in Japan studies in the Western world.

By 1890 Morse had assembled one of the best collections of Japanese ceramics in the world and had become an authority in this field, but it all began quite by chance. In the late fall of his second year in Japan he suffered from intestinal problems of nervous origin, and his Japanese doctor advised his taking daily walks to relax him. Morse did what he was told and on one of his walks in the city he discovered a ceramic plate in the form of a seashell. He then started to buy similar plates on his daily walks until he learned from a Japanese acquaintance that as ceramics they were of little value. Valuable pieces bore the name of the potter. From then on he began seriously to collect ceramics.

From the time Morse had searched for seashells as a child in Maine, collecting had been his specialty, and in his studies with Agassiz he had gained further experience. With the trained eye of a scientist Morse now began to look at ceramics and to collect, document, and categorize them systematically according to form, use, type of kiln, and geographic origin. He thus created a collection according to scientific methodology.

As in the case of seashells, his sharp observational skill was of great use. He was able to remember the slightest detail, and although he could not read Japanese, he was able to distinguish the hundreds of signature seals of the various master potters.

Morse's outgoing, cheerful character made it easy for him to engage people in conversation and thus to bring his collection systematically to completion. One day, for example, Morse visited the famous politician Okuma Shigenobu, who showed him his superb ceramics collection. Morse showed so great an interest and praised the quality of the pieces so highly that Okuma gave all of them to Morse. Similarly, for Morse as for Japanese ceramics, it was very fortunate that an expert, Ninagawa Noritane, instructed him in ceramics. Shortly before his death at the age of forty-seven, Ninagawa transmitted to him within a brief period his secret expert knowledge.

The results of Morse's work on Japanese ceramics were published in 1901 in his *Catalogue of Japanese Pottery*. As mentioned earlier, his collection is today a valuable part of the exhibit in the Boston Museum of Fine Arts.

Japanese houses, too, fascinated Morse. He carefully looked around in cities and farm and fishing villages and made sketches. He described not only the outside but also interior views, the life style in these homes, construction, and building plans. He made sketches of the *tokonoma* (alcove), kitchen, veranda, garden, and even the toilet. Peasant homes, guest houses, storage buildings, paper sliding doors, designs of the handle of the sliding walls, folding screens, windows over the sliding walls, shelf units, fireplaces, Shinto house altars, bonsai miniature trees, materials and shapes of tiles, carpenter tools, and so on—his interests were inexhaustible.

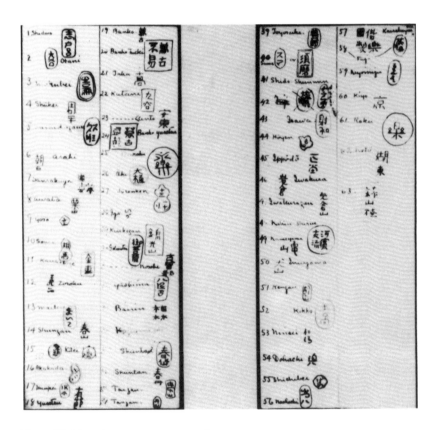

Figure 20 Morse's notes on pottery signatures.

Figure 21 Porcelain manufacture (from *Japan Day by Day*).

Figure 22 From *Japanese Homes and Their Surroundings*.

All this is contained in his book *Japanese Homes and Their Surround-ings,* which appeared in 1886 and was recently reprinted in America as a paperback. The publisher of this edition writes in his foreword that the great significance of Morse's studies lies on the one hand in their extraordinary accuracy and on the other in the fact that they came at a time when Japan had hardly yet been influenced by the West. The reprint edition is the result of the Japan boom in America. Japan studies are growing in ever greater intensity and Japanese architecture is being enthusiastically accepted.

A Japanese edition of Morse's *Japanese Homes* appeared in 1979. It is not clear whether that too followed in the wake of the Japan boom in America. Unfortunately it is very often the case—as we saw also in the *makyō* chapter—that indigenous Japan studies and the rediscovery of Japan by the Japanese require the stimulus of foreigners for a start.

Morse preserved his daily notes about his activities and experi-ences in Japan. He had planned to organize and publish them, but his zoological research and his concern to rework them thoroughly kept him from publishing them for a long time. One day a letter from another Japan scholar, William Sturgis Bigelow, reached him, encour-aging him to publish the book:

The only thing I don't like in your letter is the confession that you are still frittering away your valuable time on the lower forms of animal life, which anybody can attend to, instead of devoting it to the highest, about the manners and customs of which no one is so well qualified to speak as you. Honestly, now, isn't a Japanese a higher organism than a worm? Drop your damned Brachio-pods. They'll always be there and will inevitably be taken care of by somebody or other as the years go by, and remember that the Japanese organisms which you and I knew familiarly forty years ago are vanishing types, many of which have already disappeared completely from the face of the earth, and that men of our age are literally the last people who have seen these organisms alive. For the next generation the Japanese we knew will be as extinct as Belemnites. (E. S. Morse, *Japan Day by Day* [1917], vol. 2 preface, pp. ix–x)

Morse thereupon decided to publish and brought out his *Japan Day by Day* in two volumes in 1917. There is also a Japanese translation by Ishikawa Chiyomatsu's son Kin'ichi. In this work, Morse describes in detail the Japan of the Meiji era with the eyes of a foreigner from the time of his arrival in Japan. The contents are wide-ranging. Included are rickshaws, sumo wrestling, theater, shrines, festivals, the traditional firefighters, new year's pine tree decorations, *hagoita* (the Japanese feathered shuttlecock and battledore), women's coiffures, *ikebana* flow-er arranging, and many others. Many of these scenes no longer exist. On leafing through the pages of this book one has to admit that what Bigelow feared has come true. The Japanese who became acquainted

Figure 23 From *Japanese Homes and Their Surroundings.*

with those two belong to a species that is now extinct. Possibly the Japanese who now thrive on the Japanese islands belong to a species that survived the massive transformation of the world around them, a mutation as tough and adaptable as the brachiopods!

Morse's interests in Japan also led him to put together a comprehensive collection of ethnographic materials. There are few collections in the world equal to his, which, as mentioned before, is preserved and still exhibited in Salem's Peabody Museum, of which he was director. The collection contains small objects such as envelopes and merchants' receipts, but also Japanese wooden sandals, kimonos, kitchen utensils, Buddhist house altars, cloth carps (hung in the wind outside homes for certain festivals) and many other items—all kinds of objects of daily life in the Japan of that time. Even Japanese are surprised at the breadth and magnitude of the collection.

In addition Morse presented the various aspects of Japan and its culture in lectures and papers. He thus contributed greatly to the

Figure 24 From *Japan Day by Day.*

Figure 25 From *Japan Day by Day.*

spreading of knowledge of Japan in the West. His lecture at Ohio State University mentioned in the previous chapter is an example of this activity. To be sure, he represents the Japanese in a 1894 lecture "On the Importance of Good Manners" at Vassar College as perfect beings as far as etiquette and discipline are concerned, and one gains the impression that Morse, in his memory, idealized Japan after an absence of ten years. Be that as it may, with his lectures and publications he is said to have exerted a considerable influence on the taste and architecture, both exterior and interior, of Americans.

Japan even influenced his zoology. When after many years of absorption with Japanese ceramics he again turned to zoology, he had gained the ability to grasp the essential characteristics of animals with far simpler strokes; at least the sketches for his articles were praised for this in 1919. It is known that Japanese art and handicrafts influenced the West. This case is noteworthy, however, in showing that this influence even extended to scientific drawing.

There are numerous anecdotes about Morse and Japan. Margaretta Brooks, his assistant at the Peabody Museum, he called Otamasan, meaning "pearl" (Greek *margaron* = pearl) in the customary somewhat old-fashioned Japanese of the time. Once while traveling in a train with a guest from Japan, he suddenly started to sing a Nō-chant, to the great amazement of his fellow travelers. That he took instruction in Nō-chanting he reports in *Japan Day by Day*. He is probably the first foreigner to have studied Japanese Nō-chants with a professional—in this case, Umewaka Minoru.

Morse was advancing in age when he learned that in the great earthquake in the Kanto plain in 1923 the buildings and library of the University of Tokyo had been destroyed by fire. With the approval of his children he changed his will so that all his books were to go to the university library. After his death, sixty-nine large chests of books arrived in Tokyo. From time to time even today a book comes into one's hand with Morse's portrait on the back of its cover.

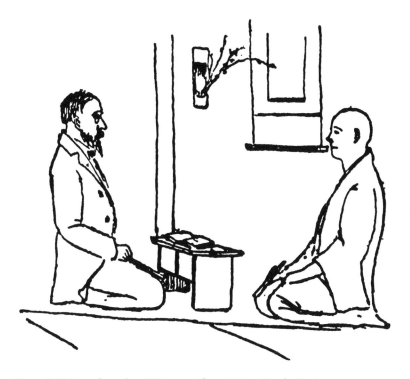

Figure 26 Morse learning Nō songs (from *Japan Day by Day*).

Science and Life

One of Morse's books is called *Mars and Its Mystery* (1906). As he himself jokingly commented, "Morse studied Mars."

His interest in astronomy too went back to his childhood. The special interest in Mars, however, arose through his friendship with Percival Lowell. Lowell, in his turn, stimulated by a lecture by Morse, became fascinated with East Asia, stayed in Japan and Korea, and wrote several books. Later he constructed an observatory at Flagstaff, Arizona, with an eighteen-inch reflector in order to extend his Mars studies, which he had started long before. Sometimes he invited Morse to join him.

Morse's friendship with Lowell and his own Mars observations at the Lowell observatory led Morse to write his Mars book. In it he compares the furrows in the Mars surface, on the one hand, with natural phenomena such as the surface cracks in the Japanese Satsuma-yaki pottery, cracks in dried mud and in asphalt, and the crater cracks on the moon and, on the other hand, with human-made net patterns as in cartographic representations of railway lines, streets, irrigation canals, and so on. From these comparisons he concludes that there must be intelligent life on Mars, thus supporting his friend Lowell, who because of his canal theory was entangled in intense controversies (see Figures 27 and 28).

In the hallway of the Peabody Museum, whose director Morse was, there was a ventilating system run by solar energy. Whether the idea came from Morse is not known, but in his own house he had an improved version of the system for heating and ventilation that worked remarkably well. He regularly measured room temperature and air exchange at various times of the day and published a report on it in 1885 in which he mentions that the same system was also used successfully in the Boston Athenaeum. Some time earlier, Morse had put together a brief report on the system, published in *Science* in 1883. Near the end of the article Professor Mendenhall, who was present when Morse presented the report, is cited as saying that he himself had seen it functioning well. The press frequently referred to "Morse's installation to harness solar energy" in sensational terms, which annoyed Morse.

One day to his consternation, on returning from a lecture tour, Morse opened his suitcase to find among his clothes a small clock that did not belong to him but rather came from the house where he had just spent the night. The ticking of the clock had bothered him the previous night; so he had put it in his suitcase and had then departed the next day, forgetting to take it out. Morse immediately sent a tele-

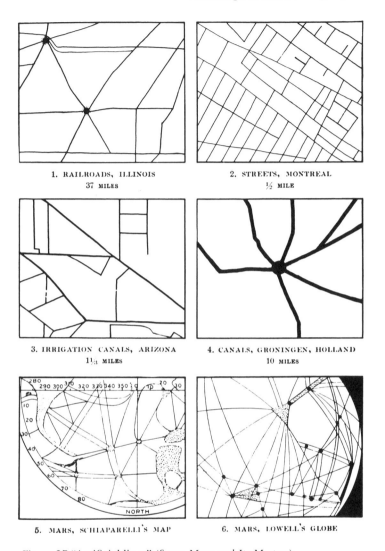

1. RAILROADS, ILLINOIS
37 MILES

2. STREETS, MONTREAL
½ MILE

3. IRRIGATION CANALS, ARIZONA
1⅓ MILES

4. CANALS, GRONINGEN, HOLLAND
10 MILES

5. MARS, SCHIAPARELLI'S MAP

6. MARS, LOWELL'S GLOBE

Figure 27 "Artificial lines" (from *Mars and Its Mystery*).

gram to the owner and returned the clock by express mail. This anecdote points to his great sensitivity to noise. He felt disturbed by the train whistles and the noise of moving trains from the nearby Salem station. Not surprisingly he lectured against noise and started a campaign against noise pollution for which he printed and distributed one of his lectures at his own cost.

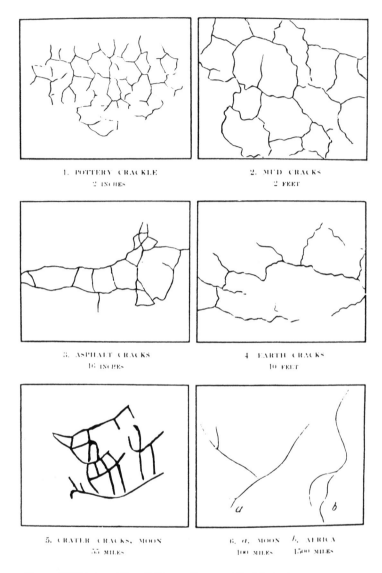

1. POTTERY CRACKLE
2 INCHES

2. MUD CRACKS
2 FEET

3. ASPHALT CRACKS
16 INCHES

4 EARTH CRACKS
10 FEET

5. CRATER CRACKS, MOON
55 MILES

6. a. MOON   b. AFRICA
100 MILES   1500 MILES

Figure 28 "Natural lines" (from *Mars and Its Mystery*).

On June 21, 1900, at graduation exercises at Worcester Institute of Technology, where Mendenhall was president at the time, he gave a lecture entitled "Can City Life Be Made Endurable?" Here too he touched on the noise problem and cited an article from the journal *Nation* that claimed that the more primitive people are, the more they love noise. Whether in joy or mourning they would shout and be noisy. I found among Morse's papers preserved in the Peabody Museum letters to an official of a railroad company discussing the noise problem. It is also said that during a banquet in his honor a friend read a poem:

St. Peter oils the hinges
To prevent Morse having twinges.

The gates of heaven were to open smoothly so that Morse would not be upset.

Morse was a skilled lecturer, and people loved his lectures. Though they might forget what he talked about, they remembered his sketches. As mentioned before, Morse was able to conjure up on the blackboard the most marvelous drawings using both hands simultaneously.

The head of the Wistar Anatomical Research Institute in Philadelphia was interested in his ambidextrous skills. Morse promised to make his brain available to the Institute when he died, and in later years he wrote to Ishikawa Chiyomatsu:

The Wistar Institute of Anatomy of Philadelphia sent a glass jar properly labelled . . . in asking for my brain which they will get when I am done with it.

After his death his brain was carefully studied at the Institute. It is still preserved in alcohol there, and I was able to see it. In a 1923 photograph in Salem, when Morse was still alive and was visited by Dr. Komai Taku and his wife, Mrs. Komai is seen in the doorway of the house holding this glass in her hand. I was able to see this photo when I visited the Komai family. The scientist and museum director Morse thus finally made even himself into an object of scientific study.

Chapter 4
Response to a New Scientific Theory: Darwinism in the Early Meiji Era

Traveling in deep snow from Fukui to Tokyo, the American science teacher William Elliot Griffis noted on January 23, 1872, half in jest:

We resume our march. . . . The tracks of boar, bear, foxes and monkeys are numerous. It is the hunter's harvest-time. Dressed carcasses are on sale in every village. I wonder how a Darwinian steak would taste? "No, thank you; no monkey for me!" is my response to an invitation to taste my ancestors. Good people, you need "science" to teach you what cannibals you are. (*The Mikado's Empire* [1886], fifth edition, p. 542)

Having just opened its doors, Japan certainly needed science, although of course not for the purpose of recognizing our ancestors but rather to participate on an equal footing with other nations. At that time the theory of evolution was spreading throughout the West. When Japan accepted Western knowledge, the theory of evolution also found its way to its shores. That theory was a new biological concept with many philosophical implications and consequences. How did the Japanese react to it? What position was taken by the foreign teachers in Japan?

First Encounters with the Theory of Evolution

From early on the theory of evolution became known in Japan in a fragmentary fashion through missionaries and foreign teachers. As a coherent system it was deliberately introduced by the American zoologist E. S. Morse, who had come to Japan in 1877 and who was the focus of Chapter 3 of this book. After giving the first three lectures on the theory of evolution on October 6 of that year, he recorded his impressions as follows:

A number of professors and their wives and from five to six hundred students were present, and nearly all of them were taking notes. It was an interesting and inspiring sight. . . . The audience seemed to be keenly interested, and it was delightful to explain the Darwinian theory without running up against theological prejudice as I often did at home. The moment I finished there was a rousing and nervous clapping of the hands which made my cheeks tingle. One of the Japanese professors told me that this was the first lecture ever given in Japan on Darwinism or evolution (*Japan Day by Day*, vol. 1, pp. 339–40)

Through Morse's lectures and their publication under the title *Dōbutsu Shinkaron (The Evolution of Animals),* which Ishikawa Chiyomatsu prepared and brought out in 1883, the theory was widely publicized in Japan. As far as the content of this book goes, the treatment of the subject is pretty crude. On comparing it with the lecture Morse gave a year earlier in America, it is noticeable how here examples from the animal and plant world are directly applied to human situations. As the title of the first lecture, "The Truth of Things Is to Be Pursued, the Teachings of Religion Are Not to Be Presupposed," makes clear, an aggressive rejection of Christianity is a prominent characteristic of Morse's lectures on the theory of evolution.

Two reasons can be advanced for this rejection: on the one hand, there was the confrontation between the theory of evolution and Christianity in the West, and, on the other, there were the premature conditions in Japan, which made it impossible there to judge the theory of evolution as a scientific theory. The former was a major problem quite generally in the Western world, but especially in Morse's case, because of an extremely strong religious atmosphere in his childhood that made him develop an aversion to it. Hence his evolution theory exhibited strongly anti-Christian tendencies. The second reason follows from the question as to what was the prerequisite for the birth of the evolution theory. A large body of information gathered from the most diverse areas of knowledge in natural history together with very careful analyses made possible the formation of the theory. In Japan, however, those areas of knowledge, which could have been the basis for the evolution theory, did not even exist. Japan had to absorb basic information and the most sophisticated theories at the same time. For these reasons it was unthinkable for the Japanese to question the scientific foundations of the theory or to make scientific judgments or criticisms about it.

In the previous year Morse in his lecture before the American Association for the Advancement of Science had pointed out that "Agassiz's earnest protest against evolution checked the too hasty acceptance of this theory among American students. . . . The results of his protest have been beneficial in one sense. They have promoted the seeking of proofs in this country." In Japan there was neither a biologist who could have taken on the role of Agassiz nor an audience that could comprehend the details or the precise significance of the theory. Morse's lectures were therefore made simple and were designed as popular science, and he made no attempt to attain a thoroughly scientific level. The content was further distorted through the conversion of the spoken lecture by translation to the printed word. Thus one can probably assume that the theory of evolution as presented in *The Evolution of Animals* was a specially tailored version prepared for his lectures in Japan and was at the same time an evolution theory on a level that the Japanese of the time were just able to understand.

For these reasons the theory of evolution was not able to assert itself in the Japan of that time as a biological theory. Many years were needed before it could establish itself within biology.

On the other hand the theory of evolution was widely disseminated and frequently discussed. To make possible a comparison with the West, representative scientific journals of that era, the English *Nature* (founded in 1869) and the American *Science* (founded in 1883), were examined. In both, the proportion of articles on evolution theory in the 1880s represented less than one percent of the total. In the corresponding Japanese scientific journal *Tōyō Gakugei Zasshi* (founded in 1881), the proportion of the total exceeds 8 percent. This amount makes clear the extraordinarily great interest of the Japanese in evolutionary theory. When the content of these articles is analyzed further, it turns out—as seen in Figure 29 (see also Figure 34 below)—that the theory of evolution is discussed more frequently in the social sciences than in the natural sciences, and this to a considerable extent.

If the books published in Japan on the biological theory of evolution are arranged chronologically, the earliest inclusion of Darwin and his theory occurs in Aoikawa Nobuchika's *Hokkyodan* (1874). Then follow *Seishu-Genshiron* (1879), a Japanese edition of Thomas Huxley's *Lectures on the Origin of Species* (London, 1862) translated by Izawa Shūji; a translation by Kōzu Senzaburo of the second edition of Darwin's *The Descent of Man* (1874) with the title *Jinsoron* (1881); Ishikawa Chiyomatsu's *Dōbutsu Shinkaron* (*The Evolution of Animals*), based on Morse's lectures (1883); an edited, expanded version of Huxley's *Lectures* by Izawa Shūji with a new title, *Shinka Genron* (1889); *Bambutsu Taika Shinsetsu* (1889), a translation of the writings of August Weissmann produced by Ishikawa Chiyomatsu, who had studied with him in Germany in 1887–1889; and another book written by Ishikawa, *Shinka*

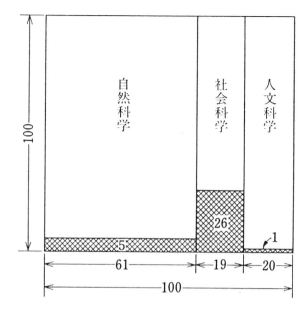

Figure 29 Numerical ratios of all articles appearing in the 1880s in the journal *Tōyō Gakugei Zasshi* according to field, with the percentage of the articles dealing with the theory of evolution in each field (words in diagram: Natural sciences, Social sciences, Humanities).

Shinron (1891). The Japanese translation of Darwin's *Origin of Species* (London, 1859) only appeared in 1896, prepared by the humanist Tachibana Sensaburō. Its title was *Seibutsu Shigen.*

By comparison, in the twelve years from 1877 to 1888 at least twenty translations of Spencer's work appeared in Japan. Considering that only four works on biological evolution theory appeared in the same timespan, the contrast becomes very clear. Such a tendency can be recognized generally in Japanese publications on evolution theory; that is, the largest fraction deals with Spencer's philosophy or social Darwinism. Only a few are devoted to the biological theory of evolution. This is characteristic of the Japanese reception of evolution theory, which was received not as a biological doctrine *per se* but either as a new interpretation of the world based on "science" or else as a simple slogan such as "victory of the superior, defeat of the inferior" or "survival of the fittest," applied mainly to social problems.

Figure 30 Title page of the Japanese edition of Darwin's *On the Origin of Species.*

The Theory of Evolution as Social Theory

As pointed out in the previous section, in the field of biology the theory of evolution hardly operated in Japan during this time. As a philosophical idea, on the other hand, it played an important role. In contrast to Europe, where the thesis of the common origin of animals and humans formed the center of discussion, the central topics in Japan were "the struggle for survival" and "natural selection." An evolution theory abbreviated to the formulas "victory of the superior, defeat of the inferior" and "survival of the fittest" was understood here as the newest scientific truth and as an unassailable basic principle and served as the basis for various isms and tenets. It was used as a weapon against Christianity and the so-called civil rights movement (*jiyu minken*) and served even for the defense of Shintoism and Buddhism. Darwin's theory of evolution presupposed a struggle for survival among individuals of the same kind; social Darwinism widened this to apply to society and the contest of countries among each other. In Japan it even served directly to justify the slogan "to enrich the country and strengthen the military" and was misused as the basis for nationalism.

A representative example would be *Jinken Shinsetsu*, the *New Theory of Human Rights*, by Katō Hiroyuki (1836–1916). Katō was president of the University of Tokyo when Morse came to Japan. He had previously produced three publications—*Rikken Seitairyaku* (*Outline of Constitutional Government*, 1868); *Shinsei Taii* (*Outline of Actual Politics*, 1870); and *Kokutai Shinron* (*New Theory of National Structure*, 1874)—and espoused in them with enthusiasm the thesis of natural rights (*tempu jinken-setsu*) or, in his own words, "the conception that we humans have been endowed with our rights by heaven." In 1879, however, he fundamentally changes his position. In November of that year and the following March he gave a lecture entitled "Against the Thesis of Natural Rights," personally prohibited the further sale of his three books, and in 1882 brought out his *New Theory of Human Rights*. Later he explained the change in his ideas:

That I fundamentally changed my ideas is connected with the fact that I read Buckle's *History of English Civilization* and recognized for the first time that metaphysics was something well-nigh absurd. It became clear to me that any statement is meaningless if not based on science, and thus I began to read Darwin's theory of evolution and the evolutionary philosophy of Spencer, Haeckel, and others. In the process my view of the universe and of life was changed from the ground up.

In the introduction of his *New Theory of Human Rights*, Katō proclaims that, just as "the disciplines that deal with nature gave up their illusions" through the heliocentric theory or the theory of evolution, so also "the disciplines dealing with the human mind such as philosophy, political science, and law" must give up their illusions. His aim is "to attack the thesis of natural rights on a scientific basis with the help of the doctrine of evolution."

Katō then explains the theory of evolution. He presents it as a theory that "elucidates the reasons why higher species arise in the struggle for survival and through natural selection in the animal and plant world." He declares this is "the eternal and unchangeable law of nature" and the "fundamental principle of all principles":

In what follows I want to designate this basic principle as the victory of the superior and the defeat of the inferior. The universe is one vast battlefield. In order to preserve and expand existence, all beings on this battlefield are engaged in constant strife and seek a decision as to victory or defeat. There is no result that is not governed by the principle of the victory of the superior. . . . We must realize that the basic principle of all principles, the victory of the superior and the defeat of the inferior, applies not only to the world of animals and plants but also with the same compelling necessity to the world of human beings. (Katō, *New Theory of Human Rights*)

Katō then explains the principle of the victory of the superior by means of examples from human society and comes to the conclusion:

In the face of these facts, there arises from the differences in human beings a continuous many-thousand-fold victory of the superior and the defeat of the inferior. This is the basic principle of all principles, an eternal and unchangeable law of nature. It is evident that neither the individual human being nor humanity as a whole has natural rights to freedom, self-government, and equality. Yet the advocates of this illusion are incapable of understanding this unarguable and clear principle of truth and busily defend natural rights. They maintain that the natural rights to freedom, self-government, and equality are

Figure 31 Title page of Katō Hiroyuki's *Jinken Shinsetsu* (*New Theory of Human Rights*). Vague and blurred, the four characters *tempu jinken* ("rights endowed by heaven", i.e., natural rights) can be seen. They are meant to indicate the illusory character of this idea.

inherent in each individual and should not be taken or violated by force. One can only laugh at this stupidity and blindness. (Ibid.)

Katō was "convinced that with the mighty sword of evolution he had destroyed with one stroke the illusion of natural human rights." Since in this lecture he abruptly connects the theory of evolution with the rejection of the doctrine of natural rights, his argument here seems rather far-fetched. Probably one was to see this as nothing more than an indication that he had simply changed his ideas.

Next Katō insists that among the races as among countries there are superior and inferior ones and that a single people also consist of superior and inferior individuals. On this basis he rejects universal suffrage. Should it happen that with a sudden change in the evolutionary process in the animal and plant realm the superior also die out, an expansion of rights and a change in society should only proceed slowly. Out of fear that his writing, based on the principle of the victory of the superior and the defeat of the inferior, would be branded as a "mischievous book that would encourage the populace to injustice and immorality" and as urging the "uprising of the lower against the upper class," Katō makes an effort to explain in conclusion that this only meant that "those predestined to win will win while those predestined to lose will lose and it is by no means an encouragement to insurrection." There was according to Katō a "good victory of the superior" where "the good wins and the weak loses" while there is also a "bad victory of the superior" in which "the bad wins and the good loses."

In this respect the argumentation all through the *New Theory of Human Rights* is confused. To say "those predetermined to win win and those predetermined to lose lose" is, as the basis for the "victory of the superior," a purely tautological assertion and one wonders how such a meaningless proposition could "in one stroke destroy" the doctrine of natural rights. For Katō there is a "good" and a "bad victory of the superior." If now, as he believes, "the right is not necessarily the superior and the superior is not necessarily the right," then those at the time superior are not necessarily the right and good. Consequently, Katō's argument that one must first acknowledge the presently superior and must avoid a sudden change to avoid exposing the same superior to extinction totally collapses.

Katō Hiroyuki's *New Theory of Human Rights* was strongly criticized by the defenders of human rights and others. Their objections can be summarized as follows:

(1) Katō prides himself on thinking that his rejection of the doctrine of human rights is a new theory. In reality that viewpoint is already out of date in scholarly circles.

(2) The doctrine of human rights, contrary to Katō's assertion, does not necessarily contradict the theory of evolution.

(3) Katō ignores the distinction between natural phenomena and human rights and reason and considers them erroneously as equally subject to the rule of "victory of the superior."

(4) Humans possess rights of two kinds, based on natural morality and on law. The rule of "victory of the superior" applies to the former, not to the latter. The human rights discussed by Katō are legal rights and thus obviously not natural ones.

(5) Natural rights become more necessary the more the world approaches the rule of "the stronger devours the weaker" and "victory of the superior."

(6) Katō's assertion that there is a "good" and a "bad" victory of the superior is "in the highest degree an incomprehensible statement."

(7) Katō asserts that the "victory of the superior" is the "law of heaven," but at the same time he fears the possibility of a victory of those whom he himself stamps as inferior—socialists, communists, and nihilists. This is a gross contradiction.

(8) The theory of evolution is "in reality an instrument for reformers." But Katō uses it to buttress his conservative theory. This is "as if, having assembled soldiers, they shoot at their own castle when ordered to attack the foe." No wonder he cannot succeed.

Just as Katō used the theory of evolution as the basis of his argument, so his opponents used it also as a "weapon." Both sides proceeded on the assumption that the theory of evolution was an established scientific truth and built their assertions on it. For Darwin the struggle for existence and natural selection were hypotheses to explain certain facts in biology such as mutation, population growth, and adaptation. In Japan, on the other hand, Darwinism was from the start an "eternal and unchangeable natural law," and it was applied without hesitation to human society. The connecting of problems of state and society with the theory of evolution was carried forward in subsequent years by both conservatives and reformers, and there developed on the basis of this theory a philosophy of life also as well as a critique of civilization, although this occurred at a later time.

The problems of "artificial selection" and "race improvement" also were given attention in connection with the theories of evolution, genetics, and eugenics. The very first article of the first issue (1881) of *Tōyō Gakugei Zasshi*, a general science journal that was important at the time, was Katō Hiroyuki's contribution, "The Possibility by Artificial Selection to Produce Gifted Human Beings." Number 29 of the same journal in 1884 contained an article, also by Katō, entitled "A Major Doubt Regarding Artificial Selection in Our Society," which was also discussed by readers in the journal. In 1884, in his *The Improvement of the Japanese Race,* Takahashi Yoshio suggested that the Japanese race could be improved through crossing with those of Europe. Katō Hiroyuki and others voiced their objection.

The leading representative of social Darwinism who sought for an evolutionary interpretation of human society was the English philosopher Herbert Spencer (1820–1903). Spencer considered society and the state as an organism and the separate individuals as cells and speculated about the mutual relations between society and the individual. From that he developed his evolutionary interpretation of the development of human society. His theory established itself in America and thence found its way to Japan. In the first chapter of this book we pointed out that Yamagawa Kenjirō had already encountered Spencer's views in *The Popular Science Monthly* during his studies in America, and it was this encounter that led him to study physics.

An even more committed follower of Spencer was Toyama Shōichi (1848–1900), who had gone to study in England in 1866 and in America the following year, later becoming professor of law at the University of Tokyo and finally its president. At the same time he was also a significant opponent of Katō Hiroyuki's *New Theory of Human Rights.*

Ernest Francisco Fenollosa (1853–1908), a teacher of philosophy at the University of Tokyo who had come to Japan in 1878, introduced Spencer's ideas to Japan. During his stay in Japan he brought out for this purpose in volume 7 of the journal *Gakugei Shirin* (1880) a three-part article in Japanese translation entitled "Setai Kaishinron" ("Theory of Social Development"). Many of Spencer's writings and essays were also translated and published and very soon books appeared in which Japanese picked up Spencer's ideas. Examples are the works of Ariga Nagao (1860–1921), whose book of 1883, *The Relation of Society and the Individual,* was joined by a three-volume set, *Theory of Evolution of Society* (1883), *Theory of Evolution of Religion* (1883), and *Theory of Evolution of Races* (1884).

At a time when the theory of evolution, as applied to society, was being disseminated throughout Japan and when the establishment of the Japanese parliament in 1890 had led the defenders of human rights to increase their disputes, Yatabe Ryōkichi (1851–1899) returned from a period of study in America and became professor of botany at the University of Tokyo, His critical views in an article with the title "On the Indignant Disputants," which appeared in the March 1886 issue of *Tōyō Gakugei Zasshi,* deserve special attention. In it we read:

Figure 32 Darwin's house.

For thirty years the English naturalist Darwin studied earthworms. . . . Although this kind of scientific research might seem ludicrous to a person of shallow learning, it did make possible many geological and physiological discoveries and it is therefore not only very valuable for science but also of enormous importance for our future. Not one of these indignant disputants would be capable of such an achievement. Darwin once remarked that Herbert Spencer's writings were excellent, yet in order to establish all his chapters scientifically would require fifteen years of research per chapter. Is this not a revealing example of Darwin's position! On studying the biographies of people who made great contributions such as the scientist Newton or the inventor Watt, one learns how such achievements come about. (Yatabe Ryōkichi, "On the Indignant Disputants," *Tōyō Gakugei Zasshi* [March 1886])

In the village of Down in southern England, barely an hour by train and bus from London, Darwin's house can still be visited in its original state. Some years ago I was there also. Near the entrance gate there was an impressive inscription: "Here Darwin worked for forty years." The thought that Darwin could do research in these pleasant surroundings—a spacious house and a magnificent garden—almost engenders a feeling of jealousy. In a corner of the garden a round stone was set into the lawn, with a sign indicating it was the "worm stone." This stone Darwin used for his earthworm studies mentioned by Yatabe. He measured to what extent the stone sank each year due to the digging of the earthworms. His study contains the apparatus with

Figure 33 The "worm stone" in Darwin's garden.

which he was able to measure the minute sinking of the stone. Considering that Darwin after his long travels on the *Beagle* had spent twenty years in research before publishing *On the Origin of Species* and that he devoted such care even to the study of earthworms, it becomes understandable that Spencer's ideas appeared to Darwin as gross bragging. Yatabe, on the other hand, who was more or less acquainted with the subject matter, must have felt very bitter and unhappy noticing the existence of these "indignant disputants" who adopted the theory of evolution overnight without a glance at the significance of continual research activity on the underlying material and who venomously argued about totally baseless topics. One could say that this piece by Yatabe describes most succinctly the core of the problem of Japan's reception of evolution theory and beyond that of the general reception by Japan of Western science and culture.

The Theory of Evolution and Christianity

The theory of evolution contradicted the Christian dogma of creation and placed human beings, seen until then as the crown of creation, in the same developmental sequence as animals. In the West this led to intense controversy. Even today there is widespread belief that the theory of evolution is irreconcilable with Christianity. If, on the other hand, the problem is viewed from a broader perspective covering many centuries, if in particular the non-Christian culture and intellectual history of Japan are compared with their Christian counterparts, then the theory that at first glance seems anti-Christian is seen as fundamentally the product of the Christian cultural realm.

The central theme of evolution theory is the transformation of species, but in order to create such a concept there first had to be the concept of a species, above all the idea of an unchanging species. This had its origin in the Judeo-Christian worldview, which designated the species as parts of God's creation. Next, to conceive of changes and development as embedded in the flow of time presupposes a concept of time flowing in one direction as well as the idea of a unique, unrepeatable progression. This was supplied by the time concept in the Judeo-Christian eschatological view of history, which understood history as a unique process running from the creation to the last day of the world. Neither of these is imaginable in a non-Jewish, non-Christian world.

Not only the theory of evolution but Western science as a whole is in many ways closely bound up with the philosophical cultural ground from which it sprang, particularly with Christianity, its most important element. Recent studies have been making this increasingly evident. The Japanese at that time, however, totally separated the two essential elements of Western culture, Christianity and science, and while one was as far as possible rejected, the other was all the more eagerly adopted. This situation arose from four particular conditions: (1) the Japanese were not acquainted with the connection of modern science with Christianity; (2) for the foreign teachers who were instructing the Japanese in the sciences, the philosophical and cultural basis of the Western world was a self-evident assumption and they therefore were not concerned about its connection with modern science; (3) with the theory of evolution the impression was strengthened that science and Christianity were irreconcilable; and (4) there is a sense of estrangement toward Christianity on the part of the Japanese that has deep historical and intellectual roots.

The theory of evolution, which attracted extraordinarily great interest, and the manner in which Morse presented it in Japan strengthened these tendencies. Thus here, in contrast to the West, the theory was accepted almost without opposition or criticism and from the beginning was even used as an intellectual weapon against Christianity. The missionaries living in Japan, the convinced Christians among the foreign teachers, and Japanese Christians were all convinced that Japan could only be modernized by adopting Christianity. How did they come to terms with this situation?

They too gave lectures; they published articles and books and developed a Christian dialectic. They dealt with the relation between Christian belief and evolution theory with replies to Morse and criticism of Katō Hiroyuki and Spencer. In the monthly magazine *Rikugō Zasshi* (founded in 1880), which was published by the Tokyo Seinenkai (Young Men's Christian Association of Tokyo) with the goal of spreading Christian doctrine, enlightenment, and research, many articles of this kind appeared. Comparing the contents of this magazine according to the various disciplines in Figure 34 with the earlier diagram for *Tōyō Gakugei Zasshi*, in Figure 29 shows that in both the proportion of articles on the theory of evolution, in all the disciplines but especially in the sciences, is decidedly large. This shows, on the one hand, how seriously the theory of evolution was taken, and witnesses, on the other hand, to the special role of Christianity in the reception of this doctrine. Over and over again *Rikugō Zasshi* published scholarly criticism of the theory of evolution as presented in Morse's lectures or in Ishikawa Chiyomatsu's adaptation in his *Evolution of Animals*. This criticism was of a kind not to be found in *Tōyō Gakugei Zasshi*. As laypersons, they could not take on the role played by Agassiz in America; the significance of this small group of Japanese and foreign Christians lay in the fact that they were the only ones who critically illumined the theory while all others accepted it without opposition. Often they did not totally reject evolution theory but attempted to harmonize it in some way with Christianity. One approach kept the two areas strictly separate, while another combined them. The most optimistic version attempting to combine the two was put forward by the missionary Guido H. F. Verbeck (1830–1898), who asserted that the law of victory of the superior applied also to religion and for that reason Christianity was spreading throughout the world as the superior religious community.

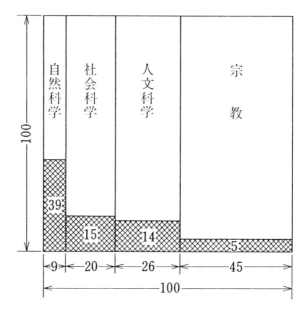

Figure 34 Numerical ratios of all articles appearing in the 1880s in the journal *Rikugō Zasshi* according to field, with the percentage of articles dealing with the theory of evolution in each field (words in diagram: Natural sciences, Social sciences, Humanities, Religion).

Among this group, John Thomas Gulick (1832–1923), who came to Japan as a missionary, takes on a special significance. Gulick was a missionary who at the same time, as a scientist, was an adherent of evolution theory. The third son of a missionary in Hawaii, he studied the *achatinellidae*, a species of land mussel, in Oahu and discovered that they varied slightly from valley to valley. This he compared with Darwin's discovery on his voyage on the *Beagle* of the differences in giant tortoises and other animals on the Galapagos Islands. Gulick always carried his huge mussel collection with him for research purposes, and thus he discovered that Japanese land mussels showed regional differences similar to those in Hawaii, although of an even more complicated kind.

The results of his life-long research he published only late in his giant work *Evolution, Racial and Habitudinal* (1905). In contrast to Darwin, who assumed that selection through the environment was the main cause of evolution, Gulick declared, using as his example the geographic distribution and the correlated differences in one species of mussel in Hawaii, that in addition to natural selection there must be

Figure 35 John Thomas Gulick (1889).

other causes for mutation and development. As a major cause he put forward the factor of isolation, that is, the "prevention of mingling." He also emphasized the complexity of hereditary processes and the effect of a new type of selection that is undertaken by the developing organisms themselves; that is, he pointed to the significance of internal as against environmental causes. The significance of Gulick's scientific work, which is being recognized again today, lies in his pointing in this direction. In 1878 and 1879 Gulick gave lectures at Dōshisha University. After discussing evolution based on selection, he presented the following argument for his claim that "selection cannot totally explain the secret of the origin of life":

(1) Selection presupposes the existence of living organisms. It is therefore not in a position fully to explain their origin.

(2) It can often be observed that organisms under the same external conditions change due to geographic separation.

(3) Sexual selection at times operates in opposition to natural selection (for instance in varying hairiness) and in this case the factor of conscious choice is stronger than natural selection.

He ended his lecture with the comment that he himself believed in evolution, but that as yet its mechanisms, laws, and causes had only been investigated to a small extent.

In a lecture about four years later (in 1883 or 1884), at the Protestant Missionary Conference on "Evolution in the Organic World," Gulick declared that the derivation of evolution from natural selection had only limited validity and that there were many other important causes. For example, a good spiritual environment could help dissolve the contradictions of the natural environment. He thus presented an interpretation of evolution that did not conflict with Christianity.

What is distinctive in Gulick's viewpoint is a "non-fatalistic self-determinism"—that is, a rejection of fatalism and determinism and the recognition of the possibility that organisms are capable of making their own decisions. Gulick emphasized that evolution theory rather than contradicting religion served to emphasize its significance. For the sake of human racial and cultural development, making the right choices was essential, and here lay the reason why religion and science had to be taken equally seriously.

This was the characteristic viewpoint of Gulick, a missionary who at the same time as a scientist defended the theory of evolution. Since, however, his sphere of activity was in Kobe and Osaka, since his main task lay in missionary work rather than in science teaching, and since he himself was a reticent person, his scientific achievements and views of evolution theory won nowhere near the influence in Japan as did those, for instance, of Morse.

The Theory of Evolution Against the Background of a Different Worldview

When the theory of evolution came into prominence in the middle of the nineteenth century, it was a major shock to the Western world. The picture of humanity as the crown of creation and the worldview associated with it seemed shaken to their foundation. By comparison, the introduction of the theory of evolution into Japan, where another religion and another view of human beings prevailed, apart from its effect on the small group of Japanese Christians was not a particularly shocking event. This comprehensive theory found entry into a culture that originally lacked a scientific basis and was used here in the form of an even more broadly understood social Darwinism. In the framework of general modernization, it served as the basic supporting scaffolding of the most varied isms and tenets, was able to spread widely, and stimulated public discussion on many topics. In the varied aspects of the Japanese acceptance of the theory of evolution, the bewilderment of the Japanese was evident as they came in contact for the first time with Western science, people, and social conditions. In the area of the biological theory of evolution, it remains clear that only very recently have the Japanese been able to answer the challenge of the West.

Chapter 5
Biology and the Buddhistic View of the Transience of Life: Oka Asajirō

Oka Asajirō was a biologist who made a name for himself as an able advocate of and commentator on the theory of evolution in the Meiji and Taishō eras. Recently there has been renewed interest in him because of his pronounced un-Japanese characteristics. His pedagogic ideas, characterized by a modern and Western educational emphasis encouraging doubt and questioning, are highly applauded. Not only his thesis of the decline and extermination of the human race but his whole attitude is said to reveal a pessimism quite untypical for a Japanese. It may be perceived, however, that Oka's writings are inconspicuously permeated with the Japanese Buddhistic view of the transience of life, and this is closely linked with his evolution theory and biology. Such a linkage with elements of tradition represents a noticeable exception in the process that introduced modern science into Japan.

This chapter focuses on Oka Asajirō. To clarify the issues, his biography and his thought-world will first be introduced, followed by a presentation of his central ideas.

Oka's Life and Thought

Oka Asajirō (1868–1944) was born in Kakezuka in the region of Iwata in Shizuoka prefecture. He completed a non-regular course in zoology at the Tokyo Imperial University in 1889, did two further years of research in the zoology department there, and then went to Germany. He spent a year studying zoology at the University of Freiburg switching to Leipzig for the next two years.

He noted that at Freiburg University "I was not particularly impressed with the teaching of the zoology professor Weissmann. That is why I switched to the University of Leipzig, where I studied for two

Figure 36 Oka Asajirō.

years under Professor [Rudolf] Leuckart. The three years of study in Germany were for me the happiest years of my life."

Oka returned to Japan with a completed doctorate and the following year (1895) went to the Yamaguchi high school. From 1897 until his retirement in 1929 he taught biology at the Tokyo Higher Normal School. After his retirement he gave lectures on the theory of evolution at this institution and at Tokyo University of Literature and Science as a part-time lecturer.

His biological research dealt with the comparative anatomy of sea squirts and leeches with the aim of clarifying "the phylogenetic relationship among the subspecies." He published over a hundred articles in Japanese and an equal number in European languages. Foreign scholars, who respected him greatly, assigned technical names that included the term "Oka" to at least seven animals.

Oka was not only a biologist but also a critical thinker and popularizer. Besides his research reports he wrote biology texts, general introductions to the field, and numerous essays. His ideas can best be seen in these introductory writings and essays, which have been compiled into the following books:

Shinkaron Kōwa (Lectures on Evolution Theory), 1904 (revised edition 1914, reprinted 1969).

Shinka to Jinsei (Evolution and Life, a collection of essays), 1906 (revised editions 1911 and 1921, reprinted 1968).

Jinrui no Kako Genzai oyobi Mirai (Past, Present and Future of Mankind), 1914 (reprinted 1968).

Seibutsugaku Kōwa (Lectures on Biology), 1916 (reprinted 1969).

Hammon to Jiyū (Anguish and Freedom, a collection of essays), 1921 (reprinted 1968).

Saru no Mure kara Kyōwakoku made (From Monkey Herd to the Republic, a collection of essays), 1926 (revised edition 1934, reprinted 1968).

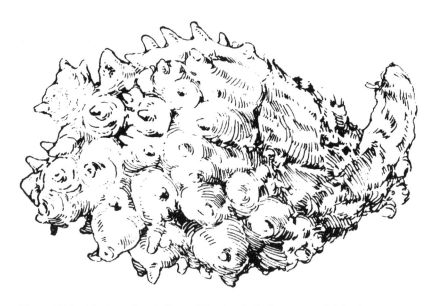

Figure 37 Ascidia (sea slime) (from Oka Asajirō's *Lectures on Biology*).

Examining these writings for their views of nature, humanity, society, education, and civilization immediately makes it clear that they are all based on Darwin's theory of evolution. Philosophy, education, social improvement, and pacifism are for Oka empty theory of no practical value if they are not grounded in the realization that the present state of development of humanity is the result of a long process originating in lower organisms. The idea that "the stronger wins" or that "the adapted organism survives" was for him a "deduction from experience" with the same truth value as the facts that "a stone falls downwards and water flows downhill." According to Oka, Darwin's evolution theory, together with his theory of natural selection that clarified these ideas, destroyed "the delusion that humans are unique in being spiritually endowed." Hence the first task is to demonstrate by means of biology and evolution theory that humans do not differ fundamentally from animals in structure, origin, function, and behavior. From this is derived the necessity to put humans on the same level as other organisms and to view them "without any preconceived ideas." This evolutionary worldview is the supporting pillar of Oka's thought system.

Education for Oka is a process that supplements or completes the generative function, and "its aim as in generative function is undoubtedly the preservation and thriving of the races." In the same way the "goal of social betterment is the preservation and prospering of the human race to which one belongs." Thus also the reason for and significance of morality and religion lie exclusively in their role in the fight for the survival of the race. According to Oka the human intellect originally was an important weapon in the struggle for survival. After a certain degree of human development, however, the intellect was also used "outside this struggle, for purposes of leisure" and thus there arose "so to say as a derailment of the intellect," as a kind of "by-product," philosophy, religion, and superstition. There is thus no harm in pursuing philosophy for pure enjoyment, in the same way that people enjoy music, painting, the game of Go, or tennis, but "it is the height of stupidity to make the great mistake of mixing conclusions from philosophy with truths about the universe and thus to ruin oneself."

Since the normal state of the world according to Oka is not peace but war, namely the struggle for existence, it is absolutely necessary to be superior in science and civilization in order to win in the battle of nations and races.

In the human struggle for survival also, that is, in the battle of nations and races, the stronger always wins. In this respect humans do not differ in the least from other organisms. It would be rash to be optimistic about one's own country in the belief that with the progress of civilization war would end and the struggle for survival would become less intense. It will be necessary always to be on guard and to see to it that this optimistic view does not spread.

On this viewpoint finally rests his plea for good scientific training and an increase in industrial productivity.

The highest aim, according to Oka, is the preservation and thriving of the race or nation, and from this comes the value of the individual:

Seen from the standpoint of the individual, his own life is of course the most important, and taking the individual as reference point, his death is the death of the universe. If on the other hand we take the race as reference, the significance of individual life is totally changed. . . . The worth of an individual's life corresponds to the amount of education devoted to him to bring him to maturity. . . . That the human being needs incomparably more education than an animal explains why human life is esteemed incomparably higher than that of an animal.

Oka further points out:

Since man is accustomed to consider his life as especially valuable, he believes by analogy that the life of other beings is also valuable and even goes so far as to praise as an especially good deed when an insect's life is spared. . . . It is based on a misunderstanding to consider the commandment not to kill as a basic ethical value as is done in the Indian religion, where one hesitates even to kill a mosquito or flea. That leads nowhere.

For the same reason "it is a real disadvantage of a race in the competition with others to have many old people in it." Regarding the abolition of the death penalty, he believed, "from the standpoint of the preservation of the race, this view is not only unfounded but definitely harmful. . . . Purely for the preservation of the race it is advantageous to eliminate without hesitation multiply-punished repeat offenders."

Oka writes further:

From the viewpoint of racial hygiene, granting medical aid to unhealthy people is not advisable, in fact probably bad.

And

For the principle of natural selection the unit of struggle is the country or the race. This must always be kept in mind when speaking of hygiene. A group's present and future should be looked at with the aim of health improvement, while the life and health of an individual is only worth improving to the extent that it is not in conflict with the overriding goal.

Thus, for Oka, social hygiene and welfare work serve no really important function.

In education, especially in science education, Oka considers it important that students develop their own doubts, think, and explore through their own motivation. This requires a class size of no more than ten to fifteen so that the teacher can concentrate intensively on each pupil. This call of Oka's is still very relevant today. However, Oka justified it not with the aim of developing each individual but rather as a means for the "survival and furtherance of the race."

In Oka's system of ideas based on evolution and natural selection, the preservation and development of the race have absolute precedence over those of the individual. Thus also the well-being of the race is the final goal of philosophy, culture, and education and their value derives from their usefulness in the achievement of that goal.

Apart from the thesis of the decline and extermination of humankind and the Buddhist idea of the transience of life, which we will discuss in the following sections, this completes our overview of the outlines of Oka's system of ideas. What he had in mind was determined through and through by the theory of evolution. By comparison with what Morse and others at first taught, Oka's evolution theory is much further developed in interpretation and mode of presentation as well as in the way he applied it. We must recognize, however, that he tried so hard to deny the sharp distinction between humans and animals, presupposed by traditional biology and Western philosophy, that he equated the two much too simply. It is, as Oka says, certainly not valid to view human beings in disregard of biology and evolution theory, but to consider humans, with Oka, almost exclusively in this way probably overlooks some important human characteristics.

In one of his essays Oka wrote: "Once we look at the logic of people today from the viewpoint of evolution theory, we must assert that it is quite wrong to assume that a conclusion is correct if it is deduced by the rules of logic. Such absolute confidence in logic is a misunderstanding on the part of those who have lost sight of the process of evolution." But does this not apply equally to Oka's own arguments based on evolution? Oka always warned against mixing facts and scientific arguments, but did he not, unknowingly, make the same mistake here?

Of course this tendency is found not only in Oka but more or less in all who based their worldview and philosophy of life on the theory of evolution. It is, so to speak, an "evolutionary monism" corresponding to the "energy monism" dominant in the same period among physicists who sought to explain everything exclusively from energy considerations. On the other hand, it needs to be recognized that with Oka the theory of evolution as mentioned already appears in much greater detail and clarity than in Morse's presentation in the early Meiji era. We must also not overlook the fact that his scientific ideas and his views on science education, when they touch central topics, contain aspects deserving considerable respect.

The Decline and Extermination of the Human Species

The most important conclusion derived by Oka from applying the evolution theory to human beings was that the human species was on the way to its own extermination. In an essay entitled "The Future of Humanity," written in November 1909 and published in January of the following year in the magazine *Chūō Kōron,* he put forward this thesis probably for the first time. (A partial translation of this article is said to have appeared in a St. Louis, Missouri, newspaper.)

At the start of the article Oka explains how he arrived at this view. Using the same approach "by which the past motions of heavenly bodies [such as Halley's comet, which was expected to appear that year] were precisely measured, their laws of motion pursued, and their movements made predictable, . . . he investigated the past changes in organisms, sought for the laws that determine the rise and fall of each species, applied them to humans, and attempted to predict their future." Of course in this case it was impossible to carry out mathematical calculations like those done in astronomy, but he was convinced that it was possible to predict "the direction humanity will take and the end-point it will reach."

The contents of this article can be summarized as follows: "In contrast to such inconspicuous and almost imperceptible mussel species such as *lingula* and *nautilus* that have existed from antiquity until today," the giant lizards of the Mesozoic or the dinosaurs of the Triassic era, which for a time had absolute superiority on earth, did not survive more than one era. Oka accounts for this surprising fact by suggesting that these animals at a certain time had characteristics that gave them an advantage in the struggle for survival. When these characteristics are too strongly expressed, however, they become a burden and prove to be a disadvantage so that they finally lead to the extermination of the species. The same could be said of humans also: "The function of brain and hands, which once established superiority over the animals and which permitted the victory of civilized people over savages, will do harm in the future. Thus the fate of humans resembles the parabolic curve of the stone thrown into the sky and falling back down: with ever-increasing speed humanity hurtles to its end."

As concrete signs of this decline and extermination Oka cites

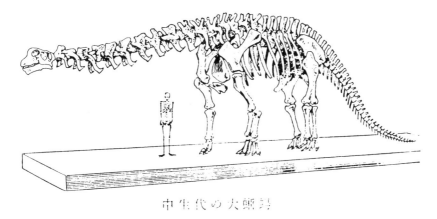

市生代の大鯏掃

Figure 38 Mesozoic giant lizard (from Oka Asajirō's *Lectures on Evolution Theory*).

moral decay and spiritual decline because of an ever-sharper disparity in wealth, a lower physical resistance due to higher levels of civilization, an increase in mental illness, suicide, and criminality, increasing insecurity and dissatisfaction stemming from the growth of knowledge, the increasing frequency of terrorism and rebellion, the decay of customs and the spread of venereal diseases arising from social problems, the dissolution of the family and harm from stimulants such as alcohol and tobacco, harm from food additives such as salicylic acid and saccharin, the increasing frequency of occupational diseases due to increasing specialization, and the lessening of a cooperative attitude because of a growing private egoism.

In 1904, the year that the Russo-Japanese war began, Oka published his *Lectures on Evolution Theory*. There the extermination of the human species is not yet mentioned. He is mainly concerned to explain evolution theory, to use it as a basis for a new interpretation of human beings, and from this viewpoint of the preservation and thriving of his own race to reorganize philosophy, religion, education, and society. The idea of the decline and extermination of humanity he developed around 1909. Thereafter he presented it repeatedly in his writings, and it formed from then on the basic motif of his view of humans and the universe. He emphasized each time that this was his own theory, one never previously put forward by anyone. The further he developed it and the more it took on substance, the more gloomy became its tone.

It seems that Oka adopted this idea in response to the contradictions, confusion, and decay of a post-war atmosphere flushed with victory. He developed it further by comparing with the world of organisms the experiences of the period after the first world war—the rise and decline of various nations, the difficulties of arms reduction and of establishing international peace, the massive advances in science and technology, political instability, labor unrest, and the economic crisis. He concluded that degenerating humanity like the accelerating fall of a stone was approaching its end. Even education and racial improvement could not prevent this falling; at most it could slow it down. Endlessly he pleaded that if possible the degeneration should be prevented, the decay slowed down, to prevent defeat at the hands of other races and countries.

An article written in 1919 and published in January 1920 in *Chūō Kōron,* entitled "The Era of Anguish," ends with the following words:

Civilization is nothing but the name for the weapon we use in battle to force other nations to fall or not to be forced by them to fall ourselves. Countries with inferior civilization are plunged into the abyss one after another. Countries with superior civilization also embroiled in conflict, slowly slide down the slope. Of course there is anguish in waking from the pleasant dream that humanity is on its way to paradise and discovering that instead it is in a situation resembling Dante's *Inferno.* This anguish of humanity is not going to lessen but will continue to the next stage, the era of despair. (Oka, "The Era of Anguish," *Chūō Kōron* [January 1920])

In an article entitled "Compassion for the Descendants," which was written in November 1923 and appeared in January 1924 in *Chūō Kōron,* he presents the later version of his theory of decline and extermination in considerable detail. Here is a summary.

Using their brains, human beings invented tools, which their hands used, and thereby they vanquished all other creatures and became lords of the earth. Thereafter only the group victorious in the struggle for survival remained. The important characteristic in this struggle is an instinctive sense for cooperative action, and normally this characteristic would have developed ever further through selection. However, since humans in the struggle for survival had gone over to the use of tools, the size of the human group in contrast to animals could increase almost without limit. The struggle for survival among large groups is less intense than that among smaller ones so that the majority of the vanquished groups survived and were able to have offspring. Thus the effect of selection weakened and was finally cancelled and the inborn sense for cooperative action began to atrophy.

Oka classified animals living in groups as having two kinds of organization: insect communities such as bees and ants that have an unusually well-developed group instinct and that are organized on the principle of equality, and class communities like the mammalian herds that presuppose the total subjugation of the individual in order to be effective in their cooperative action. With humans, since they belong to the second category, this instinct was slowly lost as natural selection was eliminated with the growth of the size of its groups. At the same time the ruling class misused its power. Demanding freedom, the lower classes rose up against their subjugation and cooperative action deteriorated further.

In this sense the history of humanity resembles a parabolic curve. When the group was still small and natural selection was still operating, the ascent occurred. With the group's excessive growth and the elimination of selection, the curve turned downwards. Present-day human beings have characteristics both from the time of ascent (typically altruism) and from the time of decline (typically egoism). The apparent contradictions in human behavior are a proof of this thesis.

Today [Oka claims] it is generally recognized that the physical degeneration of human beings began at a time when they used their brains and hands to embark on an unnatural way of living. Brain and hands, which originally helped in the victory over animals, now pull humans down both physically and mentally. It is hardly to be hoped that humans can regain the inborn sense for cooperative action that they had at the height of their development.

This summarizes Oka's article. At the end he writes: "It is the gist of my thesis that with respect to material civilization children are superior to their parents; with respect to community spirit, on the other hand, they are inferior."

This then is Oka's theory of the decline and extermination of humanity. It is not without problems, however. Imanishi Kinji criticizes Oka for causing confusion by splitting the single process of evolution into two strands, evolution and degeneration. "If natural selection is a general principle of evolution, then degeneration also should be explainable by it. By asserting instead that degeneration started because natural selection no longer operated, selection for him stops being a generally valid basic principle of evolution."

Imanishi argues further that organisms could only survive through making the environment their own. This they achieved by adaptation to their environment, thereby securing their own evolution (direct evolution). From this there arose the peaceful coexistence of all organisms in terms of habitat segregation. Oka was now attempting to explain this direct evolution by means of natural selection, which had nothing to do with it, and it was this that led him to such a strange hypothesis. He also confusedly equated individual and social evolution and jumped unduly from biological evolution to human development. In addition he tried to interpret the conduct of nations in terms of biological evolution. Imanishi's critique, however, does not invalidate Oka's commendable achievements.

The Idea of the Transience of Life

Oka's ideas about the future of humanity, as we have indicated above, are very pessimistic, but this does not necessarily mean that he was a pessimist. His central ideas probably tended to make his mood pessimistic, but from the beginning he was against pessimism. He felt it was foolish to read pessimistic philosophies or to commit suicide and warned: "Doing nothing and justifying it because of inner anguish leads even more certainly to defeat in the struggle for survival and hence to an even worse situation." He wrote this during a wave of suicides by young people who threw themselves down the Kegon waterfall in Nikkō. His *Lectures on Evolution Theory* contain the following section:

It is obvious that all is transient, but to forsake the world for that reason is a great mistake. Before the branch of a tree dies, it withers. Similarly, it is an indication of the forthcoming extermination of a race for human beings to flee from the struggle for survival because of a feeling that all is transient. To avoid the suicide of the race we should avoid a religion that has such a tendency.

All this was written before his theory of the decline and extermination of humankind was written (in November 1909), but even in the greatly revised new edition this part was left unchanged; so one can assume that his attitude toward pessimism had not altered. All along he was convinced that the world was transitory and humanity was doomed to die out. Yet to succumb to pessimism was in his eyes not an appropriate response. He would observe the events of the human world as he did the animal and plant world "without any preconceived ideas"— that is, objectively and at a distance—and stayed true to his method of observing reality, deriving his thesis from his observations and acting accordingly.

However, we have to admit that even if reality is observed objectively, its interpretation and the overview from which theory is derived is a process characterized by the individuality of the particular investigator. Oka's theory of the decline and extermination of humankind, whose uniqueness he himself emphasized repeatedly, arose of course in the same way. Important factors in the construction of the theory were Darwin's evolution and selection theories, as well as many facts from the world of nature and events from human society, but behind all this there was operating, perhaps even unknown to Oka himself, the traditional Japanese idea of the transience of life deriving from Buddhism as the conceptual framework.

Oka first mentioned this thesis of the decline and extermination of humanity in his November 1909 essay, "The Future of Humanity." He was looking, he wrote there, "for the laws determining the rise and decline of each species, applying them to humanity and attempting to predict its future." There emerged an example in human history quite similar to the flourishing and then extinction of the giant lizards of the Mesozoic and the dinosaurs of the Triassic eras, namely the splendor and the fall of the Heike family:

> Look at Japanese history. The decline of the Heike family and the rise of the Genji must not be seen as if the Heike could have developed further and could have consolidated their preeminence if the Genji had not been superior. Rather, the Heike were already showing inner signs of decline and that is the reason why the Genji could defeat them in spite of their much less advantagious position. Just as a decayed tree can be felled by a light wind, so the proud Heike went down in a short time, for their decline, threatening from within, had already weakened them when the Genji attacked from outside. . . . Their destruction had the same cause as the sudden extinction of the giant lizards and dinosaurs.

Oka ends the essay with the sentence:

> Always it has been proclaimed that the mighty will fade away. All that begins has an end—this is the law of life and death.

The fall of the Heike is a dramatic phase of Japanese history that is described in the early thirteenth-century *Heike Monogatari (The Story of the Heike Family)*, which has been loved for centuries. It describes the rise and fall of the Heike family in the so-called Gempei war (1180–1185). The first sentences of this story already proclaim the transience of everything earthly and the fall of the mighty; the whole work is permeated with the Buddhistic view of the transience of life (*mujō*). Employing the theory of evolution and trying to view humans from the point of view of biology and the theory of evolution, Oka probably unconsciously adopted this Japanese idea of the transience of life from the *Heike Monogatari* as his frame of reference, in the face of his experiences of war, confusion, and transformation in human society. From there he developed his own theory of the decline and extermination of humankind.

Oka's formulation of the "law of life and death" already quoted is found also in an old Buddhist text. In it is a sentence that we quoted earlier—"It is obvious that all is transient . . ."—and in his 1916 *Lectures on Biology* we read:

The fall of the mighty and continuous shifts and changes are eternal laws of the organic world with no exception whatsoever.

The principle of organic life expressed earlier in Japan as "victory of the superior" and "survival of the fittest" were now reformulated by Oka to read "the fall of the mighty" and "continuous shifts and changes" in terms of Buddhistic phraseology.

Seen in this way, Oka in his *Lectures on Biology* places life under the categories of "feeding, giving birth, and dying." The fact that he organized his lectures in this sequence shows that underneath lay a view of life molded by the Buddhist idea of transience of life, which supplied the conceptual framework for his theory. Moreover, Oka assumed that the rule "feeding, giving birth, and dying" holds not only for the individual but for the whole race, and he therefore entitled the last chapter "The Death of the Race," devoting the very last part of this chapter to the decline and extermination of humankind.

Oka's conception of transience of life contains elements that clearly distinguish him from Western evolutionary pessimism. He was not merely borrowing Buddhist vocabulary and concepts.

When decay and the end of humanity become unavoidable, even objectively observing scientists cannot close themselves off from feelings of sympathy for children and grandchildren. His article "Compassion for the Descendants," which he wrote in November 1923, when Japan suffered the great earthquake in the Kanto plain, concludes with a quotation from a poem from one of the oldest Japanese anthologies, the *Kokinshū*, with which Oka also expresses his own feelings:

You want to see the light of the world, little bamboo shoot? Don't you know how much sorrow it harbors?

This article was included in Oka's last collection of essays, *From Monkey Herd to the Republic*, containing pieces written between the end of 1923 and 1926. The new edition also contained an article written in November 1927 as well as one written especially for this edition. When he wrote this last one he was sixty-five years old. It is the last article he ever wrote for his book, and he had written nothing for a long time before it. Entitled "The Joys of the Unfortunate," it begins with the following words:

"If the time is not right, even Confucius has no guests and stays alone," says the beginning of the *Furai Rokubushu* [a satirical piece by Higara Gennai]. When I recently opened the book and read this phrase, I started as always to think about it and my thoughts wandered further. What situation might be meant by the statement that the time was not right? What kind of person finds himself in that state? Certainly he is dissatisfied, but does he have no joy besides? Is there perhaps a certain state of the soul that only that kind of person attains? With these useless thoughts I whiled away my time. (Oka, "The Joys of the Unfortunate")

Then he himself answers the question:

What kind of person is it for whom the time is not right? Usually he is greatly gifted but is not recognized and must content himself with a low-level position.

Oka writes as if he is totally uninvolved, but in reality is he not speaking of himself? As he wrote in his 1926 article "Failure in Examinations and Expulsion from School," his student stay in Germany represented "the happiest years of my life," while in Japan he had suffered from failure in examinations and expulsion from school. At the University of Leipzig he won his doctorate with distinction, but at the Tokyo Imperial University he had failed to complete the preliminary studies several times in a row and had to leave school. He could only stay there as a non-regular student and thus ended his studies without a degree. Highly regarded by his foreign colleagues because of his many publications and with seven different animal species named after him, Oka had to "content himself with a low-level position and is now reaching his closing years."

"The Joys of the Unfortunate" continues:

It is said that habit finally becomes second nature. On commencing by living in sullen scorn, this slowly becomes a habit and a person's wish for wealth and power is buried deep inside. Now it can happen that such a person experiences a kind of illumination and reaches a state of undemanding equanimity that is a characteristic of some of these unfortunates. (Ibid.)

The reason Oka had to give up his preliminary studies was that for two years in a row he had failed history although he had received superb grades in English, mathematics, and drawing. History and the two subjects of Chinese and composition, which also drove him to distraction, consisted mainly in senseless memorizing. However, not only in the world at large but also in the academic world only external forms and regular courses of study counted. Oka severely criticized this state of affairs in Japan as "contemptible." This appears clearly in his writings. Yet finally he reached a stage that he described as follows:

The unfortunate one despises everything. This becomes habitual. Then he unties all that binds him to this world and having no cravings he feels a wonderful lightness and can only laugh at the world around him. . . . If someone desires to attain true refinement, the quickest way to reach it is to put oneself into a wretched situation, become used to it, despise everything, and slowly break off all ties to the world. (Ibid.)

The poets Issa and Ryōkan he described as having "volunteered misfortunes" and wrote:

Whoever penetrates into the deepest secrets of this art [of misfortune] unties all that binds him to the world, avoids all eccentric conduct, and finds joy in such simple things as a basket of rice and a gourd filled with wine. (Ibid.)

As a young man he had studied biology, become a follower of Darwin's theory of evolution, and ceaselessly advocated the preservation and thriving of the race. To this end he had made suggestions for the improvement of education, and whenever he opened his mouth he had spoken of the struggle for survival and the decline and extermination of humankind. Yet in this last text of Oka's, not a trace of all that remains.

Oka seems here to have penetrated into the secret of the "Art of Misfortune," to have "untied all that bound him to the world," and to have attained a new spiritual state. Thus Oka's life, which at first glance seemed to have distanced itself so far from Japan, is in reality to a surprising extent a pilgrimage stamped with Japan's spirit.

In general, modern science and its thought forms and the traditional Japanese mode of thought exist in Japan independently side by side. The interlinking of biology and particularly evolution theory with the Buddhist view of the transience of life that occurs in Oka's thought represents a most unusual and noteworthy exception.

Chapter 6
Modern Science and the Japanese Conception of Nature: A Sketch

Normally nature study and natural science are the fields from which we obtain most of our knowledge about nature. This is a kind of knowledge that is universally valid and the laws of nature hold throughout the world. This, however, is not everything, for it is only one way of observing nature; it is only one form of knowledge concerned with only one aspect of nature.

The way an individual confronts nature not only varies from person to person, but also depends on the surrounding culture. The Japanese from ancient times have had their own conception of nature, just as Europeans have theirs. The scientific approach to nature grew out of the European conception of nature. It would certainly not have arisen in Japan.

When a Japanese lives in the West, the differences become quickly apparent. The contrasting conceptions of nature are admittedly only one of the differences, but they are an important one. This is the reason why Western science introduced from the West cannot immediately flourish here and why in applying this international science and technology special problems always surface. Not everything derives from the difference in the conception of nature, but it certainly plays an important part.

My own major field of study has been the history of science. For this purpose I lived for a time overseas and the above is the problem with which I constantly had to grapple. The subject is here called "Modern science and the Japanese conception of nature," but this really is much too large a theme. An exhaustive coverage is therefore not even attempted. The topic is not presented using historical sources and secondary literature. Here you will find only a sketch based on personal experiences and considerations.

Differing Conceptions of Nature in Daily Life

The year was 1963. I was teaching at an American university. In front of me sat thirty students who knew almost nothing about Japan. I read them the ancient story "The Promise That Was Kept" and asked them to write down their impressions to be discussed in the next class.

The story, available in an English translation by Lafcadio Hearn, comes from the collection *Ugetsu Monogatari (Stories for Rain and Moonlight)* and is entitled "The Chrysanthemum Promise." A samurai is waiting for his brother-in-law, who had promised him before embarking on distant travels that he would return on the double-nine festival, the ninth of September. On the morning of that day the samurai rises early and has food prepared and the house swept, and in the *tokonoma*, the festive alcove, he places chrysanthemums and waits for his brother-in-law. Until late in the evening he stands at the gate, but the brother-in-law does not appear. The samurai's old mother finally goes to bed but he waits on, convinced his brother-in-law will keep his promise. When it becomes very late he finally decides to go back into his house, but at that moment he hears hurried steps approaching and recognizes his brother-in-law. He immediately leads him into the house and presents him with food, but the brother-in-law will not touch the food and won't let the mother be awakened. Quietly he tells how he was arrested in the distant country and has been in prison till that day. Since he recalled the saying, "A person cannot go a thousand miles in a day but a ghost can," and since in the prison he was allowed to keep his sword, he has been able to return in time. With the words "look after mother," he vanishes. The samurai now knows that the brother-in-law has committed suicide in order to keep his promise. The next day he departs for the distant country, finally finds the guilty person, and avenges his brother-in-law's death.

I had thought that the loyalty of the samurai of old Japan who gave his life to keep his promise would make the greatest impression on all the Americans. I was totally mistaken. Almost no one wrote about that. The second most frequent comment had to do with the question whether the spirits of the dead really come home, but above all students were amazed that the samurai on the day of the expected return set out chrysanthemums to welcome his guest.

I was certainly surprised, but really they were right. On being asked what they would have done in a similar situation their answers were of the kind, "I would have driven around in my car." The story of a samurai who took his life out of loyalty, a story that the Japanese find deeply moving in Kabuki and Joruri, the puppet theater, made almost no impression. But the students reacted understandably to the remarkable refinement in the daily life of the Japanese of placing flowers in the *tokonoma* for a guest. Therein lies, as the students themselves admitted, something characteristically Japanese that does not exist in the West. It is the fundamental way of life and wisdom of the Japanese, which is rooted in nature and values its beauty.

The highly stylized Japanese art and literature also draw from this source. Japanese painting, Japanese gardens, bonsai, the tea ceremony, ikebana, Japanese poetry, the haiku, Nō and Kabuki theater, and many others are highly esteemed in the West today. All these elements have helped form the life of the Japanese, even into the details of daily living. Lafcadio Hearn, who came to Japan in the middle of the Meiji era, wrote how deeply he was impressed by the sense of beauty expressing itself even in the simple landscape picture on a sweatcloth that a rickshaw man carried on his hip.

Just as impressively fresh as was the impact that Japan made on foreigners who visited it for the first time, so deep was also the impression that the West made on the Japanese when they acquired the opportunity to travel there. This was particularly the case when communication between two regions was still limited and before television. By and large it still held in 1954, when I visited America for the first time. Only nine years had elapsed since the second world war and each side still knew little of the other so that the differences stood out all the more clearly and interest was correspondingly great.

Having come from a land where material austerity was a virtue, I was astounded to encounter disposable paper towels. When the students there saw the powdered green tea for the tea ceremony that I had brought with me, they said "Aha! Instant green tea." I explained that it had to be prepared according to very precisely prescribed rules, and while doing it one had to be formally seated on tatami mats and it could take so long that one's legs fell asleep. It seemed to me that here could clearly be seen the difference between the Japanese, who even while drinking tea don't miss enjoying nature, and Americans, totally oriented to efficiency and speed. At that time not even instant coffee was available in Japan. What a transformation has taken place in the last twenty years in Japan, where now endless instant meals and foods can be bought.

When I related this episode to F. R. Millikan, he from his side told a story with the same theme. He is a cousin of R. A. Millikan, the Nobel prizewinner in physics, had been a missionary in China, and was spending the last part of his life in retirement in America. That tale did not really take place, he exclaimed, and then told the following story. Three men—an Indian, a Chinese, and an American—came to look at Niagara Falls. The Indian immediately sensed the presence of God behind this magnificent natural phenomenon. The Chinese thought how nice it would be to build a hut next to the falls and to chat there with his friends and drink tea. The American, however, immediately asked how this massive energy could be used most effectively. Now Niagara Falls, in contrast to waterfalls in the Chinese mountains, is so noisy and sprays so much water over so large an area that the wish of the Chinese could certainly not be fulfilled, but, be that as it may, it is in any case interesting to see how different people's reactions to nature can be.

For the Chinese and Japanese, tea drinking and eating are not simply an occasion for taking in food and socializing. They must also always provide an opportunity for union with and enjoyment of nature. Especially for the tradition-conscious Japanese, good food is by

Figure 39 Cosmic system of the European Middle Ages (from Apianus' *Cosmographia*, 1539).

no means synonymous with plenty of calories or vitamins but rather with an aesthetic presentation corresponding to the particular time of year. It is even better if the dishes, the room in which the food is eaten, and the view of the garden are beautiful. The beauty here meant is, however, not of the superficial or decorative kind.

In the forms and names of traditional Japanese sweets, in contrast to the West, the frequent use of animals, plants, and other objects of nature is noticeable. The American teacher Griffis in his report on the "Place Names of Edo" of 1874 pointed out that these names often contain plant names and other terms from nature. This points to the special love of nature on the part of the Japanese.

Varying Conceptions of Nature: A Structural Comparison

The Western conception of nature, particularly the relation between nature and human beings, is most clearly visible in medieval cosmology. There the earth is the center of the universe, a system that only with Copernicus was seriously questioned. The conception of humans, nature, and the world expressed in this system has been retained from medieval times to our own. We Japanese, who are not a part of the Western world, can become acquainted with the structure of its conception of nature and the ideological framework supporting Western culture through this medieval cosmology.

The medieval picture of the universe is built on the theories of nature and the cosmos of Aristotle and Ptolemy, which were adapted by medieval theologians to the demands of Christian doctrine. It depicts the physical structure of the visible world and at the same time embodies religious truth. It is also the stage for the *Divine Comedy* of Dante Alighieri.

This universe is constructed as follows: At the center rests the earth, which is surrounded by rotating planetary spheres. The seven shells of heaven, from the earth outward, consist of the moon, Mercury, Venus, the sun, Mars, Jupiter, and Saturn. The eighth layer is the sphere of fixed stars, beyond which is the ninth heaven, which together with all the spheres inside it rotates around the earth once a day. The tenth, immovable sphere is the Empyreum, the seat of God.

Thus the universe was a bounded world with the earth as center and surrounded by the highest heaven. The firmament was further divided into two realms, the upper heaven bounded on the inside by the moon and the lower heaven directly above the earth. The upper heaven is eternal; the lower on the other hand is characterized by change and decay. Whereas the upper heaven is composed of the unchanging fifth element, the lower heaven consists of the four elements—earth, water, air, and fire—ordered from inside to outside according to their relative weight. In the innermost, the core of the earth, lies hell, the realm of Satan.

The universe and all living beings on earth are created by God, but human beings as described in Genesis are a special creation since they are made of earth and God breathed his breath into them. As creatures formed of earth, humans tend according to their nature toward the element earth and to fall into the clutches of hell at the earth's center. Since God breathed his breath into them, they possess at the same time the urge to climb from this terrestrial world into God's realm. Only to humans as chosen creatures is the task given to choose between the absolute opposites of heaven and hell, God and Satan, paradise and hell. No matter where humans may go on this earth, God hovers over them and sees them everywhere. At their feet hell is always present, and Satan endlessly provides temptations. Humans constantly find themselves in this tension and no one knows whether, following their natural instincts, they will end within the earth, that is, in hell, or whether they will be able to climb to heaven. This is the fundamental problem of human life, and it is the central theme of Dante's *Divine Comedy.*

Is there anything in Japan that could be compared to this system? It would probably be difficult to find something corresponding to the geometrically constructed universe, but it is certainly not out of place to look at the construction of the traditional Ikebana for comparison. Traditionally, an Ikebana flower arrangement also symbolizes the universe; its three main lines represent the elements of heaven, earth, and the human being. The longest branch symbolizes heaven, the middle one the human being, and the shortest the earth, and when arranged the human being stands between heaven and earth. The three branches are harmoniously arranged in such a way as to give the impression that they all originate in a single branch.

What then does a comparison with Western medieval cosmology reveal? In both, human beings stand between heaven and earth. There is, however, a great difference in the fact that in the Western system heaven and earth constitute an absolute antithesis and that human beings always find themselves caught in this tension, while in Ikebana heaven, earth, and the human being constitute a unity and in their harmonious combination form a perfect aesthetic creation. Here no antinomies reign of the kind found in the European model, but rather the human beings are part of a harmonious universe with which they are united.

Figure 40 Basic construction of Ikebana.

The Western Viewpoint and the Rise of Modern Science

God is the real source and starting point of the traditional Western conception of the world and nature. The world exists because God exists and everything in the universe is created by God. Over all this, humanity towers as a special creation because human beings are created in the image of God and they alone were the recipients of God's breath. This clearly distinguishes human beings from the rest of creation. Just as there is an absolute difference between creator and creation, so there exists also an absolute difference between the creatures touched by God's breath and the rest of creation. As God rules over the whole creation, so human beings rule over all other creatures. From this viewpoint there results finally the attitude of people in the West toward nature, which they look at objectively and make into an object for research and practical use.

The God here referred to is of course the Judaeo-Christian God. Alfred North Whitehead pointed out that this God is omnipotent and at the same time fully rational and human. God is no moody tyrant nor high-handedly inhumane. Thus what is created must also possess order and rationality, and it must be possible to recognize this order and rationality in nature. In recognizing it, one points to the deeds of God and experiences God's profound wisdom. Consequently, viewed from the standpoint of religion, this is a very meaningful activity. Reading "the book of the creatures"—that is, interpreting nature as God's creation—is a prerequisite to the understanding of the "book of the scriptures," namely the Bible. The investigation of nature is thus quite comparable with Biblical research.

Behind the analytical relationship of people in the West with nature stands the conviction that in it a certain order reigns, and its investigation and utilization were religiously founded and motivated in a decisive way. At a time when modern science was not yet in existence, and before anyone conceived of science as a profession, there were already people who, with the view of nature we have described, made possible the investigation of nature and the emergence of modern science.

Robert Hooke (1653–1703), a contemporary of Newton and a fellow Englishman, who was mainly active in the Royal Society, introduces his *Micrographia* (1665) with the following sentence:

It is the great privilege of man over the rest of creation that he is not only able to observe the workings of nature and thereby to maintain his life, but that he also has the capacity to analyze, compare, change, and improve it for various purposes.

This precisely is the conception of nature that sees the human species as a special creation and lifts it sharply from the rest of the creation, of nature. Nature is assigned a lower place and is observed objectively. Hence this is also the typical Western conception in which nature plays the role of a human tool, an object of scientific research and an object for use.

A little later Hooke even goes so far as to write:

The first human beings tasted of the forbidden fruit of knowledge and were driven out of paradise. We as their heirs can to a certain extent redress the balance in that we not only observe the not-forbidden fruit of the knowledge of nature but also taste it.

That Adam and Eve disobeyed the divine command, ate the forbidden fruit, and were driven out of paradise is a basic component of the Western conception of the world, human beings, and nature. Thus it came about that nature caused harm to human beings and they now had to gain their bread through the sweat of their brow. Hooke contended that humanity can regain through science at least in part what it lost through the disobedience of its ancestors. This was the important scientific vision of the future that Francis Bacon (1561–1626), the ideological pioneer of the Royal Society, expressed in his phrase "knowledge is power."

It is an interesting fact that at the time that Hooke wrote the introduction quoted above, John Milton (1608–1674), also in England, was writing his religious epic *Paradise Lost*. This is a convincing example of how deeply scientific research was rooted in the worldview of the time.

The relationships between human beings and nature as described here continued in the nineteenth century. As an example we quote from the Swiss-American zoologist Jean Louis Agassiz, whom we encountered earlier:

Man is the highest creature on earth yet if we don't realize that all natural characteristics of man derive from material characteristics [of lower organisms], we will not be able to understand that he possesses the possibility for the fall and for moral decline. Whether man uses his moral and intellectual characteristics that are lent him by heaven, and which distinguish him from the lower organisms for good or ill, depends on him.

Since this conception of humanity and nature dominated in the West, the theory of evolution that developed around the middle of the nineteenth century came as a considerable shock. The way the conception had to change in the face of this development E. S. Morse described as follows:

Through the all-encompassing theory of evolution, man had clearly to understand that he was only a part of the natural world, although up to then he had investigated it as a foreign entity lower than he.

The idea that nature was subservient to humanity was subjected to a major transformation by evolution theory, yet it would be premature to assume that the Western worldview has fundamentally changed.

The Japanese Outlook

The traditional Japanese conception of nature in which the human being seeks to immerse him- or herself in nature and to become one with it differs fundamentally from that of the European West. A haiku such as the following would never have been produced in Europe:

> A morning glory took my dipping bucket,
> I went to my neighbor for water.

There one would not praise the poetic sensitivity, but would criticize a poet who would bother a neighbor in the early morning merely because a plant had coiled itself around the bucket's rope.

The Japanese conception of nature produced the aesthetic culture and way of life that characterizes Japan, yet it differs fundamentally from the outlook from which sprang modern science and technology. The historical emergence of seismology is a good example. Since ancient times there have been innumerable earthquakes in Japan, yet it was not the Japanese who invented seismology. They behaved more like Kamo no Chōmei, who in his *Sketches from a Small Hermit Cell* (*Hōjōki*, 1212) enumerates natural catastrophes such as fires, earthquakes, typhoons, floods, drought, famines, and epidemics, reports of human fame and ruin, and describes the transience of the world, to find rest in the end in a lonely hut in the bosom of nature. "Nothing is better," wrote Kamo, "than to make friends solely with music and nature." Of course we notice in Kamo no Chōmei a not insignificant influence of the highly escapist and pessimistic Buddhist conception of life of his time, yet it should be noted that, in that same nature whose destructive force was a cause of this pessimism, he sought peace for his soul. This is typical of the Japanese relation to nature.

This kind of relationship to nature was not changed in its essence in the Meiji and Taishō eras. Thus the hero in Natsume Sōseki's novel *Kusamakura* (*Pillow of Grass*, 1906) seeks refuge in the mountains from the "ugly world of people." And the sociologist Shimizu Ikutarō believes that the reaction of the Japanese to the great earthquake catastrophe of 1923 is close to the position of Kamo no Chōmei. As they sat among the ruins, even those who had lost their home and their family experienced an indescribable peace on seeing the setting sun flooding the evening sky in luminous red. As Shimizu said, the Japanese saw in the "merging with beautiful nature" the "rescue from the raw nature" that destroys people, and they had "stepped into the circle in which the human being who is thrown down by the violence of nature is lifted up again by nature's beauty." From out of this circle no seismology can arise. No matter how often Japan is visited by natural catastrophes, always it is said that "The next catastrophe comes when the last one is forgotten" (Terada Torahiko). In even more extreme form one could say that no matter how fast catastrophes pile up they can never reach the speed with which the Japanese forget them. To investigate earthquakes and typhoons scientifically, to find an explanation for their cause so that a method can be found to prevent them, such a systematic process would not have taken place within Japan.

That seismology nevertheless found a foothold and developed in Japan is to be ascribed solely to the fact that the foreign teachers in Japan in the early Meiji era themselves experienced earthquakes. Observing nature objectively and seeing it as an object of research, they began, when barely confronted with this phenomenon, to investigate it in that they, for instance, developed suitable measuring instruments. They also founded the Japanese Seismological Society and its journal and laid the foundation for the development of research.

During my stay in America, a Japanese botanist who was spending some time doing research at the same university I was connected with took me for a ride in his car. On this occasion I became acutely aware how different such a ride would have been with an American. My fellow countryman did not go non-stop on the highway to our destination to return full speed, after a brief stop. Rather, he drove his old-fashioned car slowly along a curving country road. Through him for the first time I discovered that in America too there are places of great diversity and individual atmosphere. While driving he pointed to a tree in front of a farmhouse and told me how it blossomed in the spring. There we discovered a large butterfly fluttering around on the road in front of us. As I expected, my colleague stepped on the brakes and only accelerated again when the butterfly had left. When the following day I told my American colleague about this his immediate comment was: "Oh, sure, when a butterfly hits the windshield, it's a mess to remove it." I was speechless. The episode had seemed to me like a modern version of the haiku with the water bucket, but the American had totally failed to understand it. He showed absolutely no interest in the attitude toward nature and for the concern for its protection that at the time was slowly emerging in America.

One of the topics I discussed most often at the time with my Japanese colleagues in America, mainly scientists, was the question whether it was even possible for us to make a creative contribution to the development of modern science, given our traditional Japanese attitudes. The answer was uniformly no. A friend expressed it visually: if we compare the systematic development of modern science with the construction of a modern building, then the area where the uniqueness of the Japanese element could express itself would be at most in the interior architecture. I too had the feeling that a more significant contribution was unfortunately not possible.

Just by chance, however, there came into my hands at that time an essay with a much more optimistic view. It was an essay in English, "The Contribution of Japan to Modern Anthropology," by Dr. John Frisch of the University of Chicago, who was teaching at Sophia University. In it Frisch praises the recent studies of Japanese anthropologists on the social behavior of monkeys. He points out that these results have come about through a specifically Japanese research method that Western scientists until then had not emulated.

The customary Western method consists in putting out food at a given location in order to attract the monkeys and permit the researcher to observe the monkeys' activities. Previously the monkeys have to be caught and have numbers attached to them. The Japanese do not need numbers. They are able among dozens of monkeys to identify each one by its physiognomy, movements, and character. Interestingly, they also give each a name corresponding to its character or appearance. These names even appeared in the scientific publications; they are not converted into numbers. This surprised the Western scientists. It would never have occurred to them.

Frisch continues. Western scientists observe animals as if they were bacteria under the microscope. They consider them as objects. Japanese scientists, on the other hand, have a much more personal relationship with each individual monkey. They not only give each a name but know its life history and character. This process has led to valuable insights, and it is a good demonstration that the combination of Japanese elements and Western science is capable of generating noteworthy results. Frisch concludes his paper with the expectation that the Japanese, because of their special familiarity with nature, were likely in the future to make their own characteristic contribution to observing nature and solving its secrets.

This example illustrates in a graphic way that the Japanese understanding of nature is capable of making its own active contribution to the development of science. In fact, Japanese scientists since that time have amply demonstrated this capacity—for instance, in the African gorilla studies or the investigation of the way of life of the extremely rare cat species *Iriomote yamaneko* in Okinawa, which is so archaic that it has been described as a "living fossil." So far has science in Japan matured in the hundred years since its introduction. It is probably not wrong to assert that the Japanese attitude toward animals, which in these studies played a crucial role, resembles that revealed in the animal studies (*chōjū giga*) of the twelfth century.

Figure 41 Section from *Chōjū Jimbutsu Giga* (scroll of the antics of chasing animals and people) attributed to Toba Sōjō (Heian era, twelfth century). The scrolls are in the possession of the Kozanji Temple, Kyoto.

The Encounter with Environmental Destruction

On a number of occasions this book has referred to the fact that since Japan's "opening" to the West its cultural tradition has been gaining ever-wider recognition in other countries. One indication is the 1968 award of the Nobel Prize in Literature to the writer Kawabata Yasunari. In his speech entitled "Japan the Beautiful and Myself," Kawabata cited a poem of the priest Myōe:

> Winter moon, coming from the
> clouds to keep me company,
> Is the wind piercing, the snow cold?

Kawabata explained that the poet was not talking about himself but was addressing the moon and asking him if he was cold. Kawabata introduced the poem as an example of the Japanese feeling for nature.

This poem was written more than 750 years ago, but in spite of the fact that Japan since then has become an industrial nation, the Japanese feeling for nature has hardly changed. Today they sing:

> The chimneys are so tall!
> Poor moon! All enshrouded in smoke.

That is expressed in a poetic way, but in it one can discern a problem. Until recently the Japanese were quite unaware of the damage to humans and nature caused by the smoke from these chimneys. Supported by this lack of awareness and by a typical Japanese optimism, an extraordinary economic growth was welcomed and energetically pursued.

The Japanese see their relation to nature as analogous to that between mother and infant. No matter what the child does, the mother sets it right. If waste water is released into a river, nature is sure to clean it up. In earlier times this attitude worked well enough, but no change has occurred since we introduced Western technology and developed into an industrialized country. Even when cadmium and mercury are run off into a river, the assumption still is that it will clean itself up. We have learned the basic principle of chemistry, the law of conservation of mass, but we do not act accordingly. Thus pollution becomes ever more serious, the damage becomes more disastrous, and before we know it a point is reached where for the population, nature, and the whole culture there is no return.

In the countries of the West the awareness of environmental destruction began earlier. Seen from the Western perspective of nature one can say that in the West there is a clear awareness that humanity dominates nature and exploits it but also bears a direct responsibility for the consequences. This differs markedly from the Japanese conception of nature, in which there is a firm belief that all problems, including human problems, are in the end solved by nature.

A number of years ago, Lynn White, professor of history at the University of California, spoke at the Annual Meeting of the American Association for the Advancement of Science on "The Historical Roots of Our Ecologic Crisis," a paper later published in *Science* on March 10, 1967. It produced quite a shockwave in America and Europe and led to a variety of reactions. White agrees generally with what I have said so far about conceptions of nature. He too sees modern science and technology as something characteristically Western. Admittedly, various elements have been taken over from many parts of the world, especially China, but the technology that now dominates, whether used in Japan or in Nigeria, carries characteristically Western traits.

The ecological relations in which we find ourselves are determined fundamentally by our conception of nature and how we think of ourselves, that is, by our religion. Westerners recognize this unhesitatingly with respect to India, for example, but it holds true for Westerners themselves. Modern science and technology emerged from a view of nature rooted in the Judaeo-Christian tradition, and this viewpoint is thus largely responsible for the environmental destruction that modern science and technology have called forth.

After developing this theme, White declares that in the history of Christianity only one person, Francis of Assisi, who preached to the birds, broke away from this orthodox tradition. White proposes that we rethink our customary arrogance toward nature in order to pay attention to this "Franciscan" understanding of nature and nominate Francis as the patron saint of ecology.

When five years later (December 1972) I presented a lecture on "The Japanese Conception of Nature" to that year's annual meeting of the AAAS, White, whom I knew, was present. Near the end of my lecture I raised the question whether a number of Buddhists could not be added as companions to his "patron saint of ecology" and gave two examples (*Science*, January 25, 1974).

Before the monks in Southeast Asia began their early morning walks to beg for alms, they always looked at the palms of their hands because they were not allowed to go out until they could discern the lines in their hands. Otherwise they might crush small creatures in their path in the dim light. The Japanese monk Ryōkan acted similarly. During summer nights he slept under a mosquito net, not to prevent being stung by mosquitoes but to avoid squashing them inadvertently while asleep. The story goes that he always left one leg exposed outside the net so that the mosquitoes could obtain some nourishment.

That Japan in spite of this tradition has become known as "the land with the worst pollution" is due to the fact that we took on and utilized the achievements of science and technology too rapidly and unthinkingly and without concern for Japanese tradition. Seen from the viewpoint of the conception of humanity and nature, the cause is probably the lack of awareness of the self in relation to nature and the resulting lack of a sense of responsibility toward nature. To introduce modern science and technology and to profit from them requires that the customary passivity and dependence on nature be given up and that instead we face humanity and nature with an active sense of responsibility.

Chapter 7
Overall Perspectives: Tasks for Today

The Japanese first encountered Western scientific technology around 1543, when the Portuguese drifted ashore to the island of Tanega-shima. Japan acquired firearms from them. The same year, 1543, saw the publication of two epoch-making books about the macrocosm and the microcosm, Copernicus's *On the Revolutions of the Heavenly Spheres* and Vesalius's *On the Fabric of the Human Body*. From that time on, modern science grew rapidly, but because of Japan's national isolation policy her contact with Western learning was severely limited. However, this contact was maintained to a very small extent by the Dutch, who were the only Westerners allowed to carry on trade with Japan. As a base for that purpose they were given the artificial island of Dejima in the harbor of Nagasaki. Through the so-called "Dutch learning" (*rangaku*), the Japanese gained medical knowledge, in particular, and were also given intellectual stimuli that were by no means insignificant.

When Japan reopened its doors to the outside world in the middle of the nineteenth century, the West had changed considerably and its technology at the time was undergoing a breathtaking development. Science had become specialized, and technology, using the new knowledge, had realized impressive achievements in military, industrial, transportation, and communication fields. In turn this had had positive effects on science so that both technology and science had developed at an accelerating rate.

Driven by the need to create a modern state and to introduce the absolutely essential Western science and technology, Japan reached these goals in the last hundred years, but the overly hurried process did not go forward ideally in all respects. Various problems remain and strains due to too-rapid growth are evident. Far too often the new was simply superimposed on the old, which was like attempting to graft an oak onto a bamboo tree.

The Need for an Overall Perspective

Modern science and learning are activities of the human mind and intellect. To understand what this means it is not enough to learn what has been achieved in the various fields. We must also comprehend the nature of modern science, how it is produced, and what bearings it has on individuals, cultures, and societies. In the absence of such understanding, a scientist would be seriously handicapped and untrustworthy, and the average citizen helpless.

In earlier times when science had hardly any practical applications, this may not have been a serious problem. Today, however, when science coupled with modern technology has become such a powerful force, for specialists and citizens to lack a proper understanding of science is a serious cause for concern. Science can be a blind and destructive force when this understanding is lacking, and such is the world we live in. That is not to say that all problems would be solved if these insights were present, but without them the situation is very dangerous. It is clear that today both for specialists and the general public an appropriate understanding of science is essential.

Let us now return to Japan. When in the middle of the nineteenth century Japan in precipitous haste adopted Western civilization and concentrated mainly on taking over from the West an already established body of scientific knowledge, practically no attention was paid to the nature and function of this science as an intellectual activity. Moreover, since the focus was always on specialized fields, no time remained to clarify the interconnections between them and their place within the Western intellectual world and culture. This problem, present from the beginning, has become more acute in the course of time so that developments seem to have proceeded in a very unfortunate direction. Meanwhile, science has developed and fanned out further and is in the process of becoming even more powerful through its close connection with government and industry.

Since the general acceptance of Western science occurred mainly through training and education, the situation mentioned above should be rectified first in the educational field, beginning with a change in the attitude toward Western learning. Improvement is particularly needed in education in the sciences. Scientific education has admittedly by and large succeeded in producing a group of specialists in a short time. However, the majority of young people who have received this education have developed little understanding for the nature of science and have unfortunately gained an inferiority complex instead. It seems that this type of education has achieved almost nothing except to generate a strong sense of inferiority, in the sense that everyone has come to think that, since science cannot be understood anyway, it must be left to the specialists. How regrettable that education in science, instead of contributing to an understanding of modern Western thought or generating an enthusiasm for human achievements in interpreting nature, has instead become a form of intellectual torture for pupils and students and a nightmare for grownups when they recall their schooldays.

In the schools the sciences are taught in a way that shows no connection with the more than ten other subjects in the curriculum (and this situation is not confined to the sciences but is true of other subjects also). Students are given highly specialized knowledge without learning how to fit what they learn into a general picture. If this is not done by the students on their own, the school years end without the students ever having comprehended these interconnections. Then the forest cannot be seen for the trees. The compulsory and higher education courses provide only fragmentary specialized knowledge. They do not give an overview of modern thought and its achievements, which not only would be useful and interesting but in fact would be highly important. Thus school-leaving certificates, grade reports, and attendance at particular universities are considered more important than competent, broadly based knowledge. Most young people go to school not for the intellectual stimulus or broadening of knowledge but to secure the best possible career, and they choose their schools accordingly. After working their way through the laborious preparatory stage and reaching the university, they focus their interests not on their studies but on their free time. Studies are carried on, as it were, between leisure-time amusement and part-time job (even though the word "school" comes from *scola,* meaning leisure). The greater the number of students who can afford to go on to higher education, the stronger will be the tendency in these directions. This means that education no longer educates.

To create a true education there would have to be changes in a number of areas such as in the entrance examinations and the keen competition for certain educational institutions. With regard to scientific training, methods of instruction whose purpose is solely to stuff students full of partly indigestible specialized knowledge need to be given up. The focus should rather be to give students an overall perspective and understanding of science as an intellectual activity. A change must take place away from that superficial approach that gives a student the illusion of understanding science by the memorization of a number of laws and the terms for certain phenomena, thereby allowing the student to answer the assigned test questions correctly. An approach is needed through which the students make the scientific mode of thought their own and develop the ability to judge science from an overall perspective.

Modern Science as the Product of a Worldview

Modern science is valid everywhere and for everyone, yet in its origin it is the product of a Western worldview. Moreover, it did not suddenly surface in recent times but was formed on the basis of a transmission and accumulation of ancient and medieval knowledge. Thus it is deeply rooted in Western thought-forms and is tied to every aspect of that culture.

For a long time there has been a tendency to consider the Middle Ages from the standpoint of the Enlightenment as a superfluous "dark" period, to see the modern era as transcending the Middle Ages and modern science as the marvelous outcome of this transcendence. The eighteenth-century English poet Alexander Pope wrote: "Nature and Nature's laws lay hid in night: God said 'Let Newton be!' and all was light." Yet more recent studies have overturned this conception of modern science and this picture of Newton. Increasingly attention is focused on the direct links between modern science and the philosophical and religious traditions of the Middle Ages. It is being recognized how much the founders of modern science owe to the theologians and natural philosophers of the Middle Ages. In other words, modern science is being looked at in a new way as an intellectual product peculiar to Western culture.

How the modern era is related to the Middle Ages is clearly seen in the *Dialogo* (*Dialogue*) of Galileo Galilei, which is written in the form of a fictitious conversation between a representative of the conservative medieval Scholastics with a person who embodies modern science or else Galileo himself, and a brilliant, highly educated layman. It could also be called a dialogue between the Middle Ages and the modern era. From this work one can gather how modern science was formed, the nature of its historical character, and in what relation it stands to the science of the Middle Ages. We learn that modern science emerged by building on the learning of the Middle Ages and finally transcended it. The conversation takes place in the atmosphere of the Italian Renaissance. There are numerous references to painting, sculpture, and music. That the connection between the craft and academic traditions since the Middle Ages was an essential element in the birth of modern science also becomes apparent.

Galileo chose the dialogue as the format for his presentation. Behind that choice lay the view dominant since the time of the Greeks

Figure 42 Title page of Galileo's *Dialogo.*

that truth was universal and that therefore an ever greater truth can be attained through dialogue and discussion. Greek democracy too was based on this assumption. Thus the method of the dialogue as used by Galileo was not simply a method for science but was derived from fundamental conceptions that determined the political system as well. Would it be possible in Japan to carry out such a well-balanced, fruitful dialogue in politics or in the academic world? If dialogue and discussion were conducted on the basis of the conviction that truth is universal, there would be fewer cliques in politics and science and parliamentary behavior would be different also.

The following famous and much-cited paragraph is not from the *Dialogue* but from another of Galileo's writings, *The Assayer,* written in 1623

> Philosophy is written in this grand book, the universe, which stands continually open to our gaze. But the book cannot be understood unless one first learns to comprehend the language and reads the letters in which it is composed. It is written in the language of mathematics, and its characters are triangles, circles, and other geometric figures without which it is humanly impossible to understand a single word of it; without these, one wanders about in a dark labyrinth. (Stillman Drake, trans., *Discoveries and Opinions of Galileo* [New York: Doubleday and Co., 1957] pp. 237–38)

This comment shows how firmly the founders of modern science were convinced that the universe demonstrates a mathematical structure. Before modern science had been established and at a time when there were hardly any proofs of the mathematical structure of the universe, its existence was believed in nonetheless. In the stubborn and patient effort to demonstrate it, modern science was formed. Not only Galileo but also Copernicus, Kepler, Newton, and others shared the same conviction and took part in this development.

This view of the universe derives from the Greeks, especially Plato and Pythagoras. Yet the quotation from Galileo has another significance too. He compares the universe to a book. What significance did the book have for the Western world of that period? When mention was made of "the book," the Bible was meant. In church this book lay open before all eyes, but since it was written in Latin, the common people could not read it; they would have had to learn Latin first. It is probably not wrong to assume that Galileo thought of the Bible when he used the analogy. This means that the universe is a "second bible" written in the language of mathematics, as the Bible is the book of religious wisdom written in Latin, and both are "open to our gaze."

Galileo read this second bible, drafted in mathematical language, in his own way and presented the contents of what he read in the *Dialogo* and other writings. He wrote his book in Italian, which could be understood by all his compatriots, instead of following the usual practice of writing scholarly works in Latin. In certain respects this can be compared with Luther's daring decision to translate the Bible into German.

Johannes Kepler expresses even more clearly the motives and the aim to read this second bible in terms of mathematics—that is, to illuminate the construction of the universe as God's creation and thereby to proclaim his glory. In pursuit of this aim he carried out laborious calculations for decades and discovered various laws about the planets and their movements.

Prior to the discovery of the three laws that bear his name, he had found that the spheres of the six planets known in his time—Mercury, Venus, Earth, Mars, Jupiter, and Saturn—can be fitted exactly into and around the set of nested regular polyhedra, of which there are only five. The book in which he formulated his third law relating orbital

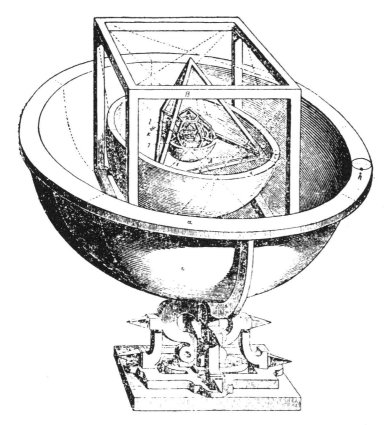

Figure 43 The arrangement of the six planets circumscribing and inscribing the five regular solids (from Kepler's *Mysterium Cosmographicum*).

radii and orbital periods he called *Harmonis Mundi*, the Harmony of the World. In it he related the planetary distances, velocities, and changes in velocities to musical scales.

Thus when he speaks of a mathematically constructed universe, Kepler meant not simply a geometric construction but also a structure in accordance with the numerical ratios of a musical scale or harmony. One is tempted to ascribe this to Kepler's mystical tendencies, but a similar approach can be found also in Newton. In his famous optics experiments on the prismatic dispersion of sunlight he asked an able assistant to mark the places where particular colors appeared. In this way he obtained seven colors from violet to red. As Newton reports in his *Opticks*, if the distance from violet to red is doubled and distances are measured from this point to each of the lines separating the colors, the distances obtained correspond exactly to the chords in an octave.

In the previous chapter we saw that the medieval conception of the cosmos is embodied in Dante's *Divine Comedy* while the modern view since Copernicus is mirrored in Milton's *Paradise Lost*. The eighth book of Milton's poem begins with a dialogue between Adam and an angel that could almost be said to reproduce the essence of Galileo's *Dialogue*. Certain parts cannot be understood without a knowledge of the history of science of the time. They are a concrete example of the fact that for an understanding of the literature of another country more is needed than a knowledge of the language. In the last chapter we also mentioned that Robert Hooke, a contemporary of Milton, in the foreword to his *Micrographia*, also talks about the loss of paradise.

Kepler's *Somnium* (*Dream*), his posthumous published work, is the first modern story of a journey to the moon, and as the work of an astronomer it takes an important place in the history of science fiction. It influenced the work of John Donne (1573–1631) and Milton. Because Kepler wrote a story about a spirit taking human beings to the moon, his mother became involved in a witch trial. The very close connection between science and literature can also be seen in the influence of the reports of members of the Royal Society on Jonathan Swift's *Gulliver's Travels* and in the influence of Newton's *Opticks* on James Thomson's verse cycle "The Seasons."

Only a few examples are presented here, but when modern science is looked at as the product of a Western worldview, and particularly its formative phase is examined, its true nature as an intellectual and spiritual human activity becomes evident. An insight can thus be obtained into its manifold and close connections with such fields as philosophy, religion, history, geography, art, music, literature, and society.

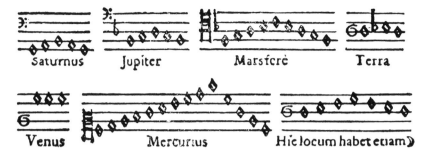

Figure 44 The relation between the greatest and least angular velocities among the six planets and the moon, with their relation to the musical scales (from Kepler's *Harmonice Mundi*).

The Confrontation with Modern Science and Technology

After its emergence in the seventeenth century, modern science went through a rapid development and achieved enormous power in the nineteenth century through its connection with modern technology. The Japanese, in the middle of the nineteenth century, were surprised by Commodore Perry's "black ships," thereby coming into direct confrontation for the first time with the scientific technology of Western civilization. It was a new civilization, characterized above all by "steam and electricity."

Perry brought a lilliputian steam locomotive and a wire telegraph machine as gifts to Japan. In Perry's report of his expedition he tells how the Japanese were amazed when he demonstrated them. In return, the Japanese staged a sumo wrestling match for the Americans. The contrast with the locomotive and telegraph in fact "was a triumphant revelation, to a partially enlightened people, of the success of science and enterprise," as Perry proudly reported.

Fukuzawa Yukichi's book *Seiyō Jijō* (*The Situation in the Western World*, 1866) shows on its cover a steam locomotive, steamboat, and telegraph together with eight Chinese characters saying "steam ferries people, electricity carries messages." In the eyes of this Japanese, who early became acquainted with the West, Western civilization was also above all characterized by "steam and electricity." Japan felt that it had to open itself to the outside world after being confronted with the power of nineteenth-century scientific technology embodied in "steam and electricity," and efforts were made to acquire this power as quickly as possible.

Figure 45 Title page of Fukuzawa Yukichi's *Seiyō Jijō* (*Conditions in the Western World*).

The foreign teachers whom Japan invited to come to Japan included the German physician Erwin Baelz (1849–1927). Baelz taught medicine at the University of Tokyo and stayed far longer in Japan than his colleagues. He loved the land and greatly admired Japanese women, one of whom he married. On November 22, 1901, a celebration was held marking the twenty-fifth anniversary of Baelz's stay in Japan, attended not only by representatives of the university but also by the minister of education and members of the German embassy. As Baelz himself reports, his speech at the celebration was "intended to be helpful and included serious suggestions for the future. I emphasized above all the need for the Japanese to acquire the spirit underlying science in order to become independently productive." He said in his speech:

Where there is light there is also shadow, and as a true warm friend of Japan I cannot overlook this shadow side. However, this festive occasion is not a suitable time to go into details.

Nevertheless, I do want to draw attention to one point since it has much to do with the development and future well-being of the university. In any event, I do not know if I will be granted another opportunity to speak not only to the leaders of science but also and especially to those on whom rests the future of science, the many students here assembled.

I have the feeling that in Japan the origin and nature of science is largely misunderstood. Science is seen as a machine that annually will produce a certain amount of work and also as one that one can simply move to another location to continue operating there. This is a mistake! Western science is not a machine but rather an organism for whose development as for all other organisms certain climatic conditions and a certain atmosphere are necessary.

Just as the earth's atmosphere is the result of an endless length of time, so also the spiritual atmosphere of the Western world is the result of thousands of years of effort by countless outstanding people to understand nature and illuminate the riddle of the world. It was a laborious road and many persons showed the way with their sweat, a road for which quite a number sacrificed their blood and their life. It is the great road of the spirit at whose beginning stand the names of Pythagoras, Aristotle, Hippocrates, and Archimedes, and whose most recent milestone carries such names as Faraday, Darwin, Helmholtz, Virchow, Pasteur, and Röntgen. This is the spirit that Europeans carry everywhere, even to the ends of the earth.

Honored guests! For the last thirty years you too have had many among you who possessed this spirit. Western countries sent you teachers who eagerly sought to transplant this spirit and to embody it in the Japanese people. Their mission, however, was often misunderstood. Originally they intended to be people who were to nurture the tree of knowledge and they proceeded accordingly, but often they were treated as if their function was to sell the fruits of science piecemeal. They wanted to sow the seedbed and hoped that from it the tree of science would independently open up and grow, a tree that with proper care would carry ever new and ever more beautiful fruit. Yet in Japan only the products of today's science were wanted. It was felt to be sufficient simply to accept the latest results and not to bother to try to understand the spirit that brought them forth." (Erwin Baelz, *Das Leben eines deutschen Arztes im erwachenden Japan* [Stuttgart: J. Engelhorns Nachfolger, 1931])

It must be admitted that Baelz was absolutely correct in his advice to change these attitudes, so as not only to acquire the results of science but to sow its seeds and care for the tree. He spoke also of the importance when with foreign teachers not only to accept the contents of the lectures but also to "glance at the intellectual laboratory from which the material that was taught originated." Unfortunately it is clear that this valuable advice has up to now seldom been heeded.

A minor example might be cited here. Galileo's discovery that the trajectory of a projectile followed one of the conic sections, the parabola, was so significant that today it would earn him a Nobel Prize. If the procedure leading him to this discovery is traced again today, the nature of modern science can be elucidated particularly clearly. However, since "parabola" was translated into a newly coined Japanese term in the mid-nineteenth century and was represented by Chinese characters that literally mean "curve of projectile," every elementary student today takes it for granted that anything thrown will follow a parabolic path. They do not first ask what path a thrown object will take and certainly are not surprised to find that the path is parabolic. The result is simply accepted without so much as a "glance at the intellectual laboratory" from which the information came. Probably they are not even aware of the existence and necessity of such a laboratory. Such improvised terminology in Japan has contributed to the quick but undigested absorption of Western science.

The same could probably also be said even of the earth's motion as proposed by Copernicus. Today no one any longer asserts that the earth is stationary. That, however, is mainly to avoid being laughed at. There are very few who think to any extent about the reasons why the earth should or should not move with the aim of coming to a reasoned conclusion. There are even fewer who will follow and become excited about the intellectual adventure of Copernicus, who on the meager basis of the knowledge of his time boldly asserted five hundred years ago, in the face of all common sense, that the earth moves. Who today is still inspired by it? Yet the root of modern science lies here, and it is probably impossible to grasp the nature of the achievement without following the process that was carried on in this "intellectual laboratory."

But is it not true that many of us who were confronted with modern science and technology, who accepted it and used it, still do not understand its nature or the "intellectual laboratory" where it was created? How widely has it been realized that its roots are in the Western worldview and that it is inseparably related to all areas of this civilization?

Concluding Comments

The pediatrician Matsuda Michio, who has been seriously concerned about Japan's problems connected with the introduction and rapid development of modern scientific knowledge, commented in 1966:

> The consequences of such a problematic development are concentrated on the weakest members of the population. The brunt of industrialization as practiced by the Meiji government was borne by the peasant girls. Hundreds and thousands of young girls became workers in textile mills and died of tuberculosis. Guns and battleships were paid for with the profits made from their silk and cotton fabrics.

Japan's transformation into an industrialized country continued after the second world war. Roads were built for industry, factories were constructed at the seaside, and from day to day the number of cars increased.

Where today are concentrated the strains produced by Japan's industrialization? With the younger children for they are the weakest.

Japan has the highest accident rate for small children. Small children die on the streets. They run after their balls rolling from the house to the road and are run over by cars.

Small children drown in rivers. Bulldozers flatten the hills on which they play. The open spaces where they chased each other were covered with concrete buildings. Where now could they play except at the river banks?

Small children play hide and seek in old refrigerators and are asphyxiated. Since no playgrounds are left it is natural that they go to unfenced factory areas.

It is easy for husbands to say, "Take good care of the children!" but what can the wife do?

Children's accidents result from the total historical development that led to Japan's industrialization. It is too strong an opponent for a mother wanting to protect her child.

If mothers wanted to protest they would not know where to find the enemy. If they took their complaint to the town hall they would be politely received and would depart with farewells and the matter would end with nothing but their photographs appearing in the papers next day.

Are children's deaths in accidents therefore a "necessary cost" for the modernization of Japan?

To defend children's human rights in the face of the profit motive and lack of sensitivity dominant in the world today is a task a father must take on who loves his children. (*Oyaji tai Kodomo [Fathers versus Children]*, 1966)

When I was writing this last chapter in 1975, there had been practically no instances where Japanese people on their own had protested against chemical damage in agriculture or the dangers of pharmaceutical products or in favor of control of automotive pollution. Before the appropriate offices and ministries react, the matter must first make the headlines overseas and be picked up by Japanese journalism or else, as has happened recently, the seriously affected citizens must take up the initiative. In this area, too, the Japanese follow the West; their sense of inferiority still remains.

This state of affairs indicates a lack of spiritual independence and genuine respect for human dignity, a lack originating from an absence of a value system that can provide a basis for a sound worldview.

The foreign teachers who came to Japan at the beginning of the Meiji era to teach scientific technology first saw the magnificent Mount Fuji on shipboard as they approached Tokyo Bay. After their arrival they climbed it, measured its altitude, and carried out gravity measurements and meteorological observations on its peak. There was also a teacher in Tokyo who looked for Mount Fuji daily and counted the days it was visible. This number has now been cut almost in half. The reason, of course, is atmospheric pollution. The story is told of a foreigner who recently visited Japan and several times travelled past the base of Fuji on the bullet train without seeing it until he finally asked if it was still there. Fuji still exists, but it is a mountain of waste paper and empty cans. The trees at its base are as good as dead because of exhaust fumes, and Suruga Bay directly below it is an ocean of chemical slime.

Our ancestors sang of the beauty of nature that they "never tired of looking at." Little of this, however, is left in Japan now.

From now on, we must have our eyes open to genuine values and must cope with modern science from an unswerving worldview and an overall perspective. I would be most happy if through its analysis of historical examples and its critical observations this book would provide stimuli in these directions. A genuine reform can probably only be achieved on the basis of historical insight.

Epilogue to the English Edition

Fourteen years have passed since this book was first published in Japan. In November 1987, I was walking in front of the National Archives, in Washington, D.C., with pleasant recollections. It was here in 1954 that I was able to find some treasured documents concerning the late nineteenth-century American science teachers in Japan. This time I was drawn to an inscription on the pedestal of a seated statue. It read, "WHAT IS PAST IS PROLOGUE." Balancing it was another statue whose inscription read, "STUDY THE PAST." Immediately I felt that this was the purpose of these Archives and the adjacent magnificent historical museums, and it was this awareness that the Japanese people had been badly lacking.

The Japanese people have made great strides moving forward but have cared little for the past. They have been keenly concerned about the *main text* but neglected to read the *prologue*. In order somehow to fill this gap, I wrote the present book of historical surveys and case studies. They describe the meeting of the Japanese people with Western science since the mid-nineteenth century.

In May 1989, I was surprised to learn that the chapter of the present book dealing with the Japanese magic mirror has turned out to be something of a direct *prologue* to recent Japanese precision technology for inspecting finely polished surfaces such as those of silicon wafers to detect minute surface irregularities. Generally speaking, however, the *prologue* is still utterly neglected, and the most recent decade has seen a further incredible development of Japanese technological industry that has caused severe economic friction, environmental disruption, and dehumanization.

Since the producing and adopting of modern science are deeply related to intellectual attitudes and more basically to one's worldview as discussed in this book, it may be difficult for most Westerners living in the West, and for most Japanese people living and working only within

Japan, to be aware of the depth of this problem. Two articles seem to address this issue: "Can Science Be Properly Done by a Japanese Also?" *Shinzen* (January 1976): 52–58, and "Questioning Japanese Creativity," a feature article in *Kagaku Asahi* (August 1987): 51–75.

After years of teaching and conducting research as professor of physics at the Technische Universität, Munich, Morinaga Haruhiko became aware of the striking differences between the Western and Japanese cultures and the consequences that these differences have had upon the essential nature of thought in these cultures. He realized that science is a uniquely Western product and questioned whether science can be properly done by a Japanese. This was similar to the questions, "Can English be properly spoken by a Japanese?" or "Can haiku be properly composed by a Westerner?" All such questions arise from the awareness of cultural differences. These differences may not be distinctly perceived by those who have remained in their own country and have been immersed in their cultural environment, as Morinaga himself pointed out in his article.

One basic cultural difference he clearly noticed was "the lack of an absolute" among the Japanese people. This seems to have led to remarkable differences in various phases of cultural life, such as logic, language, the view of nature, and human relations, which according to Morinaga must have put Japanese people at a distinct disadvantage in scientific activities.

In certain cases Japanese researchers, like the physicist Tukawa Hideki and the anthropologists referred to in Chapter 6 of the present book made original contributions under the special influence of Oriental thought or owing to uniquely intimate relations with nature. On the whole, however, we have to admit that the methodology of science in the broadest sense (including the concept of knowledge, mental attitude, method of inquiry, and so on) has been essentially of Western culture, and it has been almost unheard of in the Japanese cultural and social tradition.

The feature article in *Kagaku Asahi* consists of interviews with six internationally active Japanese scientists, including Professor Tonegawa Susumu of the Massachusetts Institute of Technology, who was later to win the Nobel Prize. To be creative, they pointed out, Japanese researchers had to acquire several traits that were foreign to them. These include a strong individuality (because originality presupposes strong individuality) and the practice of free exchange of opinions and personnel, which up to now have been largely discouraged in Japan due to its traditional social structure of seniority and factionalism.

Tonegawa especially emphasized that the Japanese people should realize that "science has essentially been a Western practice" and that "the social and cultural basis that has encouraged creative thinking in science has been closely connected to Western civilization." He also mentioned that he had received a number of research grants in the past that required no specified outcomes. He felt that the reason that this sort of thing was not found in Japan was related to different concepts of the provenance of knowledge. In Japan, knowledge had been considered as analogous to private property and not the common property of all humanity, to be shared by everybody. This situation is not unrelated to the previously mentioned "lack of an absolute."

It was over a century ago that Japan opened itself extensively to Western science and technology. This was the time when the so-called institutionalization of science was going on in Western countries. Although the Japanese people successfully absorbed the most advanced forms of these institutional aspects of modern science, they have had little concern for the non-institutional aspects of the *prologue.*

Now is the time when the Japanese people need to pause and read the *prologue,* to reflect upon the historical background of Western culture where modern science and its allied technology were produced and also to reflect on the Japanese cultural background itself into which these powerful foreign components came to be introduced.

Bibliography

Chapter 1

Watanabe Masao. "Kenjiro Yamagawa and the Sheffield Scientific School of Yale University: An Early Contact in the Field of Science Between Japan and the United States of America." *Publications of the Institute for Comparative Studies of Culture Affiliated to Tokyo Women's Christian College* 8 (1959): 127–63 and 170–71, plus 8 plates. (Japanese with English summary.)

———. "Kenjiro Yamagawa and the Sheffield Scientific School of Yale University." *Kagakushi Kenkyu (Journal of History of Science, Japan)* 53 (1960): 9–13. (Japanese.)

———. "Kenjiro Yamagawa as a Physicist." *Kagakushi Kenkyu (Journal of History of Science, Japan)* 57 (1961): 22–28. (Japanese with English summary.)

———. *Bunkashi niokeru Kindai Kagaku (Science in the History of Modern Culture)*, pp. 110–45, 315–17, and 335–37. Tokyo, 1963. (Japanese.)

Yamagawa Kenjirō. *Danshaku Yamagawa Sensei Ikō (The Writings of Baron Yamagawa)*. Damshaku Yamagawa Sensei Kinenkai (the Commemoration Committee of Baron Yamagawa). Tokyo, 1937. (Japanese.)

———. *Danshaku Yamagawa Sensei Den (A Life of Baron Yamagawa)*. Damshaku Yamagawa Sensei Kinenkai (the Commemoration Committee of Baron Yamagawa). Tokyo, 1939. (Japanese.)

Chapter 2

Watanabe Masao. "The Scientific Studies of Magic Mirrors in Meiji Japan." *Kagakushi Kenkyu (Journal of History of Science, Japan)* 61 (1962): 25–29. (Japanese with English summary.)

———. *Bunkashi niokeru Kindai Kagaku (Science in the History of Modern Culture)*, pp. 89–109, 314–15, and 334–35. Tokyo, 1963. (Japanese.)

———. "The Magic Mirror as Studied in Japan During the Meiji Period." *Japanese Studies in the History of Science* 3 (1964): 48–59. (Japanese with English summary.)

———. "Recent Studies of Japanese Magic Mirrors in America." *Kagakushi Kenkyu (Journal of History of Science, Japan)* 73 (1965): 30–31. (Japanese.)

———. "The Japanese Magic Mirror: An Object of Art and of Scientific Study." *Archives of the Chinese Art Society of America* 19 (1965): 45–51. (English.)

Watanabe Masao and Kuwata Fumiko. "The Science of Magic Mirrors in Japan During the Meiji Era." *Essays and Studies, Tokyo Women's Christian College* 12, no. 1 (1961): 1–21. (Japanese with English summary.)

Chapter 3

Gulick, Addison. *Evolutionist and Missionary John Thomas Gulick.* Chicago, 1932. (English.)

Morse, Edward S. *Dōbutsu Shinkaron (Animal Evolution).* Tokyo, 1883. (Japanese.)

——. *Japanese Homes and Their Surroundings.* Boston, 1885; New York, 1961. (English.)

——. *Mars and Its Mystery.* Boston, 1906. (English.)

——. *Japan Day by Day*, 2 vols. Boston and New York, 1917. (English.)

Ogawa Mariko and Watanabe Masao. "E. S. Morse and Zoology." *Iden (Heredity)* 29, no. 5 (1975): 77–83; 29, no. 6 (1975): 43–48; 29, no. 8 (1975): 53–60. (Japanese.)

——. "Shell Mounds of Omori as Discovered and Studied by E. S. Morse." *Seibutsugaku-shi Kenkyu (Japanese Journal of the History of Biology)* 29 (1976): 1–12. (Japanese.)

Watanabe Masao. "On Edward Sylvester Morse." *Kagakushi Kenkyu (Journal of History of Science, Japan)* 84 (1967): 161–66. (Japanese with English summary.)

——. "Additional Notes on E. S. Morse." *Seibutsugaku-shi Kenkyu (Japanese Journal of the History of Biology)* 13 (1967): 59–62. (Japanese.)

——. "E. S. Morse: Zoology, Japanology, and Other Subjects." *Meiji Taisho no Gakusha-tachi (Scholars of the Meiji and Taisho Eras),* pp. 29–66. Tokyo, 1978. (Japanese.)

——. *Oyatoi Beikokujin Kagaku Kyōshi (Science Across the Pacific),* pp. 197–273, 455–85, and 525–27. Tokyo, 1976. (Japanese.)

——. "John Thomas Gulick: American Evolutionist and Missionary in Japan." *Kagakushi Kenkyu (Journal of History of Science, Japan)* 77 (1966): 10–15. (Japanese with English summary.)

——. "John Thomas Gulick: American Evolutionist and Missionary in Japan." *Japanese Studies in the History of Science* 5 (1966): 140–49. (English.)

Chapter 4

Watanabe Masao. "Japan's Response to Darwinism." *Proceedings of the Seventh Joint Meeting of the Visiting Scholars Association of China, Korea, Japan, Harvard-Yenching Institute, Harvard University,* pp. 18–24. Cambridge, Mass., 1969. (English.)

——. "Darwinism in Japan in the Late Nineteenth Century." *Actes, XIIᵉ Congrés International d'Histoire des Sciences, Paris, 1968,* 11 (1971): 149–54. (English.)

——. "Darwinism in Japan in the Early Meiji Period." *Seiyō no Shōgeki to Nihon (Western Impact and Modern Japan),* pp. 83–107. Tokyo, 1973. (Japanese.)

Watanabe Masao and Ose Yōko. "General Academic Trend and the Theory of Evolution in Late Nineteenth Century Japan: A Statistical Analysis of Contemporary Periodicals." *Japanese Studies in the History of Science* 7 (1968): 129–42. (English.)

——. "The Trend of Academic Journals and the Theory of Evolution in the

Early Meiji Period." *Kagakushi Kenkyu (Journal of History of Science, Japan)* 88 (1968): 186–93. (Japanese with English summary.)

Chapter 5

Oka Asajirō. *Oka Asajirō Chosaku-shu (Works of Oka Asajirō)*. 6 vols. Tokyo, 1968–1969.
Watanabe Masao. "Oka Asajirō, Biology, and the Transience of Life." *Kagakushi Kenkyu (Journal of History of Science, Japan)* 107 (1973): 114–21. (Japanese with English summary.)

Chapter 6

Frisch, John. "Japan's Contribution to Modern Anthropology." *Studies in Japanese Culture*, pp. 225–44. Tokyo, 1963. (English.)
Ienaga Saburō. *Nihon Shisōshi niokeru Shūkyōteki Shizenkan no Tenkai (Religious Views of Nature in the History of Japanese Thought)*. Tokyo, 1944. (Japanese.)
Shimizu Ikutarō. *Nihonteki Narumono (In the Japanese Manner)*. Tokyo, 1968. (Japanese.)
Terada Torahiko. "Japanese View of Nature." *Terada Torahiko Zenshū (Works of Terada Torahiko)*, vol. 5, pp. 569–608. Tokyo, 1950. (Japanese.)
Watanabe Masao. "The Conception of Nature in Japanese Culture." *Science* 183 (Jan. 25, 1974): 279–82. (English.)
———. "Rescuing People from 'Traffic Congestion.'" *Shizen eno Kyōmei (Resonance in Nature)*, vol. 2: *Nihon no Hito to Kankyō tono Tsunagari (Reflections on the Relationships Between Humans and their Environment)*, ed. Kurosaka Miwako, pp. 15–33. Tokyo, 1989. (Japanese.)
White, Lynn, Jr. "The Historical Roots of Our Ecologic Crisis." *Science* 155 (March 10, 1967): 1203–1207. (English.)

Chapter 7

Baelz, Erwin. *Awakening Japan: The Diary of a German Doctor*, pp. 148–51, ed. Toku Baelz. Translated from the German by E. and C. Paul. Bloomington, Ind., and London, 1974. (English.)
Matsuda Michio. *Oyaji tai Kodomo (Fathers Versus Children)*, pp. 124–26. Tokyo, 1966.

Epilogue

Bartholomew, James R. *The Formation of Science in Japan*. New Haven and London: Yale University Press, 1989.
"Questioning Japanese Creativity." (A Feature Article). *Kagaku Asahi* (Aug. 1987): 51–75. (Japanese.)
Morinaga Haruhiko. "Can Science Be Properly Done by a Japanese?" *Shizen* (Jan. 1976): 52–58. (Japanese.)
Watanabe Masao. "Physics in Japan Since the Meiji Period: A Reconsideration and a Comparison with Australia's Experience." *Kōza Kagaku-shi (Discourse on the History of Science)*, vol. 4: *Nihon Kagaku-shi no Shatei (History of Science in Japan)*, ed. Itō Shuntarō and Murakami Yōichirō, pp. 145–73. Tokyo, 1989. (Japanese.)

Index

This book was set in Baskerville and Eras typefaces. Baskerville was designed by John Baskerville at his private press in Birmingham, England, in the eighteenth century. The first typeface to depart from oldstyle typeface design, Baskerville has more variation between thick and thin strokes. In an effort to insure that the thick and thin strokes of his typeface reproduced well on paper, John Baskerville developed the first wove paper, the surface of which was much smoother than the laid paper of the time. The development of wove paper was partly responsible for the introduction of typefaces classified as modern, which have even more contrast between thick and thin strokes.

Eras was designed in 1969 by Studio Hollenstein in Paris for the Wagner Typefoundry. A contemporary script-like version of a sans-serif typeface, the letters of Eras have a monotone stroke and are slightly inclined.

Printed on acid-free paper.